54 **Topics in Current Chemistry**

Fortschritte der chemischen Forschung

W0042540

Triplet States I

Springer-Verlag
Berlin Heidelberg GmbH 1975

This series presents critical reviews of the present position and future trends in modern chemical research. It is addressed to all research and industrial chemists who wish to keep abreast of advances in their subject.

As a rule, contributions are specially commissioned. The editors and publishers will, however, always be pleased to receive suggestions and supplementary information. Papers are accepted for "Topics in Current Chemistry" in either German or English.

ISBN 978-3-662-15986-6   ISBN 978-3-540-37394-0 (eBook)
DOI 10.1007/978-3-540-37394-0

Library of Congress Cataloging in Publication Data. Main entry under title: Triplet states. (Topics in current chemistry; 54—55). Bibliography: p. Includes index. CONTENTS: 1. Devaquet, A. Quantum-mechanical calculations of the potential energy surfaces of triplet states. Ipaktschi, J., Dauben, W. G., and Lodder, G. Photochemistry of B,γ-unsaturated ketones. Maki, A. H. and Zuclich, J. A. Protein triplet states. [etc.]. 1. Excited state chemistry—Addresses, essays, lectures. 2. Triplet state—Addresses, essays, lectures. I. Series.
QDl.F58 vol. 54—55 [QD461.5] 540′.8s [541′.24] 75—1466 ISBN 0-387-07107-5 (v. l)

Contents

Quantum-Mechanical Calculations of the Potential Energy Surfaces of Triplet States

Dr. Alain Devaquet

Laboratoire de Chimie Théorique*, Université de Paris-Sud, Centre d'Orsay, Orsay, France

Contents

---

* The Laboratoire de Chimie Théorique is also part of the Laboratoire de Physico-Chimie des Rayonnements associated with the CNRS.

1

Introduction

Over the last 15 years quantum-mechanical methods — mainly "ab-initio ' SCF—CI methods — have been developed to enable the vertically excited states of molecules to be described and the experimentally measured Franck-Condon excitation energies to be reproduced. In the case of conjugated molecules, the main target of such calculations, a reasonable description of the singlet and triplet $n\pi^*$ and of the triplet $\pi\pi^*$ states has been achieved. However, the $\pi\pi^*$ singlet state still remains the villain of the piece. A more critical test of any method for calculating electronic spectral data consists in comparing the potential energy curves of excited states with those inferred experimentally. Before we go on to discuss the methods used to calculate potential energy surfaces of triplet states, we will briefly summarize the insights they have recently brought in our comprehension of the intimate nature of the triplet state.

In a model system where two electrons are forced to occupy two orthogonal orbitals the triplet state is more stable than the associated singlet state (Hund's rule of maximum multiplicity). The conventional interpretation invokes the Pauli principle, arguing that electrons with parallel spins (triplet state) tend to avoid each other and therefore repel each other to a lesser degree than electrons with anti-parallel spins (singlet state). Let us consider, for example, the case of the carbon atom [1]: when the *same orbitals* are used for the different terms (^3P, ^1D and ^1S) within the one-configuration approximation, the kinetic and nuclear attraction energies are the same. The interelectronic repulsion energy is smaller for the triplet ^3P state than for the two singlet ^1D and ^1S states (see Table 1).

Table 1. Energy components (in au) for the ^3P, ^1D and ^1S states of carbon atom (taken from Ref. [1])

Electronic terms		Total energy	Kinetic energy	Nuclear attraction energy	Electronic repulsion energy
Optimized	^3P	−32.622	37.622	−88.106	12.861
basis set for	^1D	−37.556	37.622	−88.106	12.927
the ^3P state	^1S	−37.457	37.622	−88.106	13.026
Separately	^3P	−37.622	37.622	−88.106	12.861
optimized	^1D	−37.557	37.557	−87.963	12.849
basis sets	^1S	−37.461	37.461	−87.746	12.824

This tends to support the conventional scheme. However, the observation that in some atomic systems the interelectronic repulsion energy is higher in the higher-multiplicity terms corresponding to a given configuration than in the lower-multiplicity terms has disproved this interpretation [2-5]. If, in the case of the carbon atom, one now introduces the *independent optimization of the orbitals* for each term, the interelectronic energy indeed appears to be higher in the ^3P state than in the two singlet states (Table 1). This seems to be due to the fact that the inner $1s$ and $2s$ orbitals show a slight expansion in the ^3P term with respect to the ^1D and ^1S terms, whereas the outer $2p$ orbitals are more contracted in the lowest-energy ^3P term. This stronger interelectronic repulsion is more than compensated for by a stronger nuclear attraction energy (Table 1). To summarize these findings, one might say that, whereas interelectronic repulsion is the decisive factor in the ordering of the levels at the zeroth-order stage when the same orbitals are used for the different terms, the nuclear attraction appears to be responsible for the ordering at the first-order stage when the orbitals are separately optimized.

Our presentation of the wave-mechanical schemes used to determine the potential energy surfaces of triplet states is divided into two parts. In the first, the theoretical framework of the most popular methods is outlined and, to illustrate and compare them, the improvements which result at each step in the description of the excited triplet states of formaldehyde are briefly discussed. In the second part, a number of photophysical and photochemical problems will be analyzed on the basis of calculated potential energy surfaces.

I. "Ab-initio" Description of Triplet States

Before discussing the elaborate "ab-initio" methods now at our disposal, let us recall that less sophisticated tools are also available. These are not dealt with here, both because they have already been the subject of extensive reviews and because they have not yet reached the state of refinement that would permit their systematic and reliable use. Some of the results will, however, be mentioned in the second part of this article. These tools may be classified in three groups.

1. Completely empirical treatments, in which the potential energy surfaces of singlet and triplet states are represented as a function of potential parameters fitted to the available experimental information (equilibrium geometries, vibrational frequencies), have had a considerable success for molecules for which a *localized* electronic description is applicable [6-10]. In the case of conjugated molecules, the important *delocalization* of the π electrons introduces difficulties in such treatments [11] and the two following approaches appear preferable:

2. The first approach uses semi-empirical procedures (EHT, PPP, CNDO, INDO, MINDO) [12–16]. Many calculations of this type have been performed and remarkable results have been obtained in the evaluation of ground-state properties [12]. These methods have also been tested in the study of triplet-state potential energy surfaces (tetramethylene [17], retinal [18–19], acrolein [20], benzophenone [21], barrelene [22], propynal and formic acid [23], acetone [24] formaldehyde [25]). The MINDO/2 method seems particularly promising [26–27].

3. The second approach [28] assumes the separability of σ and π electrons. The σ electrons are represented by empirical potential functions (as in the completely empirical treatments) whereas the π electrons are treated at the semi-empirical level (a Pariser-Parr-Pople treatment including nearest-neighbour orbital overlap). Such a scheme has not yet been extensively used but looks very interesting since a single set of parameters is introduced to represent all of the molecular properties considered (atomization and ionization energies, electronic excitation energies, equilibrium geometries and vibrational frequencies of the ground and excited states). All bond length and bond angle variations are also taken into account.

The "ab-initio" SCF methods employed in the study of excited states may be classified in three groups depending on the nature of the molecular orbitals (MO's) used to build the excited-state wave function.

1. In a N-electron system the Hartree-Fock (HF) calculation on the closed-shell ground state will generate $N/2$ doubly occupied orbitals and many vacant virtual orbitals, the number of which is a function of the basis set chosen for the calculation. In the first group of methods all these ground-state MO's (GSMO's) — occupied and virtual — are used without alteration to construct the excited-state wave function. It has, however, been shown that an electron in a virtual GSMO feels the full interaction of the N electrons. Hence these orbitals are not suited for describing the excited states of the system where only N-1 electrons are available to interact with the excited electron [29]. The two other groups of methods are aimed at solving this major handicap.

2. A partial solution is to permit the virtual orbitals to adjust themselves within the subspace of vacant GSMO's in such a way as to take account of the hole created upon excitation. The occupied GSMO's may be used either without modification or by being also allowed to rearrange within their subspace, to describe the doubly occupied MO's of the excited state. These methods might be tentatively dubbed "improved ground-state molecular orbitals — IGSMO — methods".

3. In the third approach the excited wave-functions are determined without reference to the ground-state MO's. A special open-shell HF hamiltonian is built up which takes into account the field of the singly occupied MO's. However, these excited-state MO's (ESMO's) and the corresponding excited-state wavefunction are usually non-orthogonal (except by symmetry) to the GSMO's and the ground-state wave function, respectively, so that difficulties arise in the CI treatments where the values of operators between non-orthogonal functions have to be evaluated.

Where the available theoretical results allow, all the methods to be discussed are illustrated by the same example: the $n\pi^*$ triplet state of the formaldehyde molecule. In its ground state formaldehyde is a planar molecule of electronic configuration [30]:

$$(\text{core}) \ (1b_2)^2 \ (5a_1)^2 \ (1b_1)^2 \ (2b_2)^2 \ .$$

The $1b_1$ and $2b_2$ orbitals are the familiar π and n MO's, respectively. The lowest-lying empty orbital is the π^* $(2b_1)$MO (see Ref. [31]) for a representation of these MO's). The lowest $^3A_2$ triplet state of the molecule results from the promotion of an electron from the $n(2b_2)$ orbital to the π^* $(2b_1)$ orbital. Although we focus our attention mainly on the $^3A_2$ state, it is appropriate to make comparisons with two other states: the $^1A_2$ *singlet* state, which has the *same* electronic configuration, and another *triplet* state — the $^3A_1(\pi\pi^*)$ state — which has a different electronic configuration. The experimental characteristics[32–39] of the two $n\pi^*$ states are summarized in Table 2 together with the corresponding properties of the ground state [40–41]. The $^3A_1(\pi\pi^*)$ state has not been detected experimentally.

Table 2. Experimental equilibrium geometries and (0,0) transition energies of the $n\pi^*$ singlet and triplet (A_2) states of formaldehyde

Property	Ground state $^1A_1$	$^3A_2$	$^1A_2$
CO bond length R (au)	2.2825 [40,41]	2.470 [35]	2.504 [34, 35]
Out-of-plane bending angle θ (degree)	0°	35.6° [35] 35° [32]	26.9° [35] 20.5° [34] 31° [32]
Band origin and range of the transition (eV) [32]	—	3.12 (3.12–3.44)	3.50 (3.50—5.39)

I.1. GSMO Methods

Let us first briefly recall how these GSMO's are obtained [42]. The energy of a Slater determinant (where a is the usual antisymmetrizer)

$$\Psi = a \{\psi_1 \ldots \psi_N\} \tag{1}$$

built on the *spin orbitals* ψ_i is:

$$E_0 = <H> = \sum_{i=1}^{N} <\psi_i \mid h \mid \psi_i> + \sum_{j>i}^{N} (\mathscr{J}_{ij} - \mathscr{K}_{ij}) . \tag{2}$$

h contains the one-electron operators (kinetic energy and nuclear attraction energy). $\mathscr{J}_{ij}$ and $\mathscr{K}_{ij}$ are the usual Coulomb and exchange integrals between the spin orbitals ψ_i and ψ_j. The best ψ_i function, obtained by requiring the energy $<H>$ to be stationary under variation of the spinorbital ψ_i, is a solution of the eigenvalue equation

$$H_i \psi_i = \{h + \sum_{j \neq i} (\mathscr{J}_j - \mathscr{K}_j)\} \psi_i = \varepsilon_i \psi_i . \tag{3}$$

The operator H_i contains the Coulomb and exchange operators, including spin, and has a set of eigenfunctions, one of which is selected to be occupied in (1); the neglected eigenfunctions are called "virtual" orbitals. This system of N simultaneous equations is transformed into a single eigenvalue problem by constructing a unique operator H^{HF}:

$$H^{HF} = h + \sum_{j} (\mathscr{J}_j - \mathscr{K}_j) . \tag{4}$$

Since $(\mathscr{J}_i - \mathscr{K}_i) \psi_i = 0$, we have

$$H^{HF} \psi_i = H_i \psi_i = \varepsilon_i \psi_i . \tag{5}$$

The N occupied spin orbitals are now all eigenfunctions of the *same* operator H^{HF}. The last step is to express the spin orbital ψ_i as the product of a spatial orbital ϕ_i and a spin factor, α or β, and to integrate over the spin variables. The (spatial) HF hamiltonian of a singlet ($m_s = 0$) closed-shell ground state is then:

$$H^{HF} = h + \sum_{j}^{N/2} (2 J_j - K_j) . \tag{6}$$

The electron i therefore feels the *exact* nuclear attraction field and an *average* interelectronic field due to the other electrons $2J(i)-K(i)$, which is the best monoelectronic approximation of the real interelectronic field. The molecular orbital (GSMO) ϕ_i may be interpreted as the state of an electron i moving *independently* in the average field of the others. That each of the electrons may be considered independently is the direct consequence of the monoelectronic nature of H^{HF} (and not of the fact that the total wave function has been written as a Slater determinant). In the case of a singlet closed-shell ground state, the HF method yields the best possible wave function that can be given an independent-particle interpretation. However, it has to be borne in mind that the GSMO's must be adapted to the irreducible representations of the spatial symmetry group of the molecule and, as such, are delocalized over the whole system. This situation conflicts with the familiar picture of bonding in molecules which assumes, except for the π electrons in conjugated molecules, that the MO's are localized either near one center (inner shells, non-bonding lone pairs) or two centers (bonding and antibonding MO's). The attempts to alleviate this contradiction by transforming the delocalized MO's into localized ones have shown that the new localized MO's cannot be given an independent-particle interpretation [43-46] (but see Ref. [47]).

After these conceptual remarks we must emphasize two practical characteristics of the HF method which are of relevance to the present subject. Since we shall consider excited-states energies, we have to remember that, despite the high accuracy of the total energy (99% for H_2 [48], He_2 [49], N_2 [50], LiH [51]), the error is of the same order as even the largest energies of interest (binding and excitation energies). Since we shall also consider potential energy surfaces, and in particular dissociation paths, we have to be aware of the major drawback of the HF treatment of closed shells, that the dissociation of molecules is usually predicted incorrectly. For example, for H_2 at $R = \infty$, the HF wave function predicts the same probability for the two possible reaction paths (heterolytic $H^+ + H^-$ and homolytic $H. + H.$ dissociations). The error in energy at $R = \infty$ in this particular case is 7.74 eV [52]. This improper behavior of the closed-shell HF wave function at infinity complicates the study of chemical reactions because to correct it necessitates configuration-interaction calculations (but see Ref. [47]).

The simplest way to describe a triplet excited state with the use of GSMO's is to limit its wave function to a single configuration. The triplet resulting from the excitation of an electron from the occupied (spatial) orbital ϕ_i to the empty virtual orbital ϕ_l will be written [13]:

$$\Phi\,(^3i \to l) = \frac{1}{\sqrt{2}}\,\{a\,(\phi_i\,\bar{\phi}_l) - a\,(\phi_l\,\bar{\phi}_i)\}\,, \qquad E = E_0 + \varepsilon_l - \varepsilon_i - J_{il} \qquad (7)$$

7

As shown in Table 3, such a single configuration treatment gives a reasonably good equilibrium geometry for the $^3A_2$ state [53] (whereas the $^1A_2$ state is found to have its minimum energy for the planar molecule, in contrast to experimental findings) but a vertical excitation energy which is roughly 1 eV too high. Even though the error in the $^3A_2$ state is far less dramatic than that in the $^1A_2$ state, a single configuration treatment does not account for the fact that the charge distribution of a given MO differs from state to state and cannot be expected to give reliable potential-energy surfaces. The obvious remedy is to formulate the N-electron wave function of the excited state Φ_K as a linear combination of several configurations Ψ_l:

$$\Phi_K = \sum_l c_{Kl} \Psi_l, \tag{8}$$

where the Ψ_l's are themselves antisymmetrized products of molecular spin orbitals

$$\Psi_l = a \{\psi_1^l \ldots \psi_N^l\}. \tag{9}$$

The spatial part of each ψ_j^l, ϕ_j^l, is expanded in terms of the fundamental set of basis functions (Slater- or Gauss-type functions) g_m

$$\phi_j^l = \sum_m e_{mj}^l \cdot g_m. \tag{10}$$

The wave function Φ_K is determined by a variational minimization (including the orthogonality restraints) of the energy expectation value:

$$<H> = <\Phi_K | H | \Phi_K> \tag{11}$$

with respect to all or part of the available parameters (c_{Kl} and e_{mj}^l). Once the basis set g_m is chosen, several options are available for the variational procedure [30]:

1. the complete set of configurations Ψ_l derivable from the set of ϕ_j's, $\{\phi_j\}$, obtained by a linear combination of the set $\{g_m\}$ of g_m's is in cluded. In that case the only remaining degrees of freedom are the coefficients c_{Kl}, since $\{\phi_j\}$ spans the same space as $\{g_m\}$. This procedure is intractable with a large basis set.

2. If the set of Ψ_l's is truncated by restricting the occupancy of the ϕ_j's, then the optimization of Φ_K is obtained by variation of both the c_{Kl}'s and the e_{mj}^l's. This strategy has been employed [54-59] but is intractable for large basis sets or molecules of chemical interest.

3. Finally, it is useful to consider the adequacy of a set of configurations Ψ_l based on a set of MO's $\{\phi_j\}$ determined by some more restrictive formulation. We could, for example, define a core of GSMO's assumed to be unperturbed in all ground and excited states of interest; certain high-lying virtual MO's might also be excluded. A second alternative is to transform the virtual set of orbitals to enhance the convergence of the configuration interaction treatment, but this we will not deal with here [30]. Another approach (to be considered later on) is the use of MO's determined from a triplet state open-shell SCF calculation (ESMO's); after imposing core and virtual orbital restrictions, we use these ESMO's to construct both triplet and singlet states corresponding to the same spatial orbital excitation. In any case, the choice of the unperturbed core and high-lying orbitals or, more generally, the choice of the Ψ_l's always appears to be crucial. It has to be practical yet still capable of giving a suitable representation of the low-lying excited states.

    This choice may first be made on the basis of personal judgment and experience. In their study of formaldehyde [53] Buenker and Peyerimhoff allocate the GSMO's to three groups: (a) five core orbitals ($1a_1$, $2a_1$, $3a_1$, $4a_1$ and $1b_2$), which remain doubly occupied; (b) seven valence orbitals ($5a_1$, $1b_1$, $2b_2$ all doubly occupied in the ground state, and $2b_1$, $6a_1$, $3b_2$, $3b_1$, the lowest-lying virtual orbitals), which will be allowed to have variable occupancy; (c) all other virtual MO's, which remain always vacant. The CI treatment is then carried at two levels. The first one includes all single and double excitations. Some selected triple and quadruple excitations are added at the second level but this has little effect (at least in the case of formaldehyde; in other systems such configurations have an important influence [60]). If we look first at the vertical transition energies, it is clear that the energy lowering of the $^3A_2$ state upon CI is slightly smaller than the concomitant lowering of the ground state. The transition energy of $^3A_2$ is actually somewhat greater after CI than before. More striking is the fact that the CI (GSMO) treatment is not able to predict the pyramidal equilibrium geometry of the A_2 states (see Table 3). The single-configuration treatment predicts at least the correct bent geometry of the triplet state. The CI (GSMO) method at the level discussed here is not capable of compensating for the inadequacy of the GSMO's in the representation of the $n\pi^*$ excited states. Its scope needs to be greatly expanded.

    The selection of the configurations Ψ_l for the CI may also be carried out using the "points system" of Morokuma and Konishi [61] in which each MO is assigned a "point" which depends on its energy and on its potential importance to the properties to be calculated. The next step is to characterize each excited configuration Ψ_l with a point that is the sum of points of all the MO's involved in the excitation process that generates Ψ_l from the ground state. All configurations with a total of points below an arbitrary chosen limit are included in the CI. In the case of formaldehyde [62] the orbitals $1b_2$, $5a_1$, $1b_1$, $2b_2$, $6a_1$ and $3b_2$ were assigned the points 1,1,1,0,0 and 0, respectively, and the maximum number of points allowed for any configuration was 2. The results of this treatment are summarized in Table 3. The pyramidal structures of both the singlet and triplet A_2 states are reasonably well reproduced

Table 3. Treatment of the $n\pi^*(A_2)$, $\pi\pi^*(A_1)$ excited states of formaldehyde using various CI methods (Buenker-Peyerimhoff [53], Morokuma-Hayes [62], Whitten-Hackmeyer [30]) based on the ground-state molecular orbitals (GSMO's)

Property	States	Single config- uration B.P.	CI(GSMO) (1.4 exci- tations) B.P.	CI(GSMO) "point" method M.H.	CI(GSMO) "split- basis" set W.H.	CI(GSMO) "split- extended" W.H.
CO bond length R (au)	$^3A_2$	2.451	2.61	2.58		
	$^1A_2$	2.449	2.67	2.58		
	$^3A_1$	–	>2.8			
Out-of-plane bending angle θ (degree)	$^3A_2$	25°	0°	22°		
	$^1A_2$	0°	0°	27°		
	$^3A_1$	–	0°	–		
Vertical excitation energies (eV)	$^3A_2$	4.10	4.21	2.56	3.45	3.51
	$^1A_2$	4.78	4.71	2.92	3.84	3.93
	$^3A_1$	–	5.67	–	5.82	5.77

but the vertical excitation energies are now 0.7 eV too low. Due to the greater number of configurations included in the CI (186 triplet configurations for $^3A_2$, as compared with 100 in Buenker and Peyerimhoff's treatment) and also to the selection of these configurations (though the selection criterion is rather empirical) the inadequacy of the GSMO's is partly compensated.

Much more rigorous is the selection scheme devised by Whitten and Hackmayer [30] that expands the scope of the CI so that the goal of a full CI can be approximately attained within practical limits, thereby de-emphasizing the role of the GSMO basis set. Their treatment generates the configurations relevant to the CI according to the approximate magnitude of their interaction with the state of interest, as opposed to more familiar schemes which are organized around certain types of excitations (all single, or all single and double excitations [60]...).

a) A set of M parent configurations $\{\Psi_k^{(1)}\}$ is chosen which contains the configurations expected to be the most important contributors to the state of interest.

b) A second set of configurations $\{\Psi_k^{(2)}\}$ is generated by performing single and double excitations from each parent $\Psi_k^{(1)}$. The $\Psi_k^{(2)}$'s are in addition subject to a perturbation-like threshold condition

$$\left| \frac{|< \Psi_k^{(2)} | H | \Psi_j^{(1)} >|^2}{< \Psi_k^{(2)} | H | \Psi_k^{(2)} > - < \Psi_j^{(1)} | H | \Psi_j^{(1)} >} \right| > \delta \qquad (12)$$

for at least one $j = 1,2 \ldots M$. This condition takes into account the magnitude of both the hamiltonian matrix element between $\Psi_k^{(2)}$ and $\Psi_j^{(1)}$ and the energy gap separating these two configurations.

c) The combined set $\{\Psi_k^{(1)} \ U \ \Psi_k^{(2)}\}$ is completed so as to include all configurations excluded by the threshold condition that are necessary to obtain precise eigenfunctions of the spin—angular momentum operator S^2. The total wave function becomes

$$\Phi = \sum_{i=1}^{2} \sum_{k} c_{ki} \ \Psi_k^{(i)} . \qquad (13)$$

The H matrix is then diagonalized to provide the usual eigenfunctions Φ_j and eigenvalues E_j.

d) The M' lowest energy CI wave functions will now play the role of the parent set $\{\Psi_k^{(1)}\}$. Step b) is repeated with the *entire* set of wave functions used in evaluating the quantities involved in (12). The new configurations $\Psi_k^{(2)}$, plus those contained in the M' lowest-energy wave functions and the missing m_s components, form the basis for the final diagonalization process.

This scheme has been tested by Whitten and Hackmeyer on the formaldehyde molecule [30], using a "split-basis" set of Gaussian atomic orbitals. The Gaussian basis sets commonly used have a fixed-group basis [63]. This means that, say, a pπ orbital is represented by two pπ basis functions (short- and long-range functions) with coefficient ratios *constrained* to give atomic p orbitals. Such a basis set may be made more flexible in two ways: (a) by determining the coefficients of the short- and long-range functions *independently* at the SCF level, and (b) by adding more diffuse functions having n greater than 2 (3pπ basis functions, for example). A basis set improved following (a) is a "split-basis" set. At this split-basis set level the vertical excitation energies of the $^3A_2$, $^1A_2$ and $^3A_1$ states are 3.45, 3.84 and 5.82 eV, respectively (see Table 3). If this basis set is further extended following (b) by including

$3p\pi$ orbitals on C and O and a $3p$ (b_2 symmetry) orbital on O (to improve the description of the π orbitals and the n-oxygen lone pair, respectively) the values become slightly higher for the A_2 states and slightly lower for the $^3A_1$ state. For the first time in this study of the GSMO methods the vertical excitation energies appear to be within the range of the experimental values (the excited-state equilibrium geometries have not been determined). To compensate for the inadequacy of the GSMO's, the CI procedure must include approximately 200 configurations. Moreover, these configurations have to be specially selected, since they are those which have an interaction with the states of interest greater than $\delta = 2.10^{-4}$. Thus, a reasonably good result has been achieved with inappropriate tools (the GSMO's) but a great deal of effort. As the following two parts show, less work is necessary if better tools are developed.

I.2. IGSMO Methods

In the case of a N-electron closed-shell, the self-consistent field to be felt by one electron arises from N-1 partners. The formal removal of $\mathscr{J}_i - \mathscr{K}_i$ from H^{HF} (4) to give H_i(3) effects this adjustment. For the energy, this adjustment is introduced by the relation:

$$\mathscr{J}_{ii} - \mathscr{K}_{ii} = 0 \qquad (14)$$

The interelectronic energy of an electron in orbital i with two paired electrons in orbital j consists of two parts: J_{ij} for the different-spin interaction and $J_{ij} - K_{ij}$ for the same-spin interaction, which together give $2J_{ij} - K_{ij}$. Within the orbital i only J_{ii} should appear but this term, due to relation (14), may be replaced by $2J_{ii} - K_{ii}$. It is important to realize that this self-adjustment occurs only for occupied orbitals — thanks to the property $(J_i - K_i)\,\psi_i = 0$ — but not for virtual orbitals since for those the operator $2J_i - K_i$ is present, and an electron in a virtual orbital feels the full interaction of N electrons. For this reason it is often said that virtual-orbital solutions of (5) are appropriate for (N + 1)-electron systems [64]. It would be natural to use an operator like (3) to obtain appropriate virtual orbitals for N-electron systems. This has been done by Kelly in his extensive perturbation calculation of Be [29-65], by Hunt and Goddard in their calculation of the excited states of H_2O [66], and by Lefebvre-Brion et al. (frozen-core approximation) [67]. Goddard's method will serve to illustrate this general type of treatment.

Let us suppose that in the determinant (1) ψ_i is simply replaced by ψ_l. By again requiring the energy to be stationary under small changes of the spin orbital ψ_l (with the added condition that all the molecular orbitals must be orthogonal), we obtain:

$$H_i \, \psi_l = \varepsilon_l \, \psi_l \, . \tag{15}$$

This means that the variationally correct ψ_l is a virtual orbital of H_i from the ground-state calculation but, because of the self-terms in H^{HF}, is not a virtual orbital of H^{HF}. After integrating over the spin coordinates, the optimum spatial molecular orbital ϕ_l, in the case where we start with a closed-shell wave function, is a solution of

$$H_i^{EX} \, \phi_l = \{h + \sum_{j \neq i}^{N/2} (2 \, J_j - K_j) + J_i\} \, \phi_l$$
$$= \varepsilon_{ll} \, \phi_l + \sum_{j}^{N/2} \varepsilon_{jl} \, \phi_j \, . \tag{16}$$

when the spin projection operator is taken as $m_s = 0$. In the case of a triplet state with $m_s = 1$, we would obtain in similar fashion

$$H_i^{EX} \, \phi_l = \{h + \sum_{j \neq i}^{N/2} (2 J_j - K_j) + J_i - K_i\} \, \phi_l \, . \tag{17}$$

(The off-diagonal Lagrange multipliers ε_{jl} in (16) insure that ϕ_l is orthogonal to the other occupied orbitals j). Since the exact singlet and triplet wave functions are given by the two combinations [13]

$$\frac{1}{\sqrt{2}} \{a \, (\phi_i \, \bar{\phi}_l) \pm a \, (\phi_l \, \bar{\phi}_i)\} \tag{18}$$

the true H_i^{EX} operator is:

$$H_i^{EX} = h + \sum_{j \neq i} (2 J_j - K_j) + J_i \pm K_i \, . \tag{19}$$

In (18) and (19) the plus and minus signs refer to the singlet and triplet excited states, respectively. The solutions of (19) will be called "improved virtual orbitals" (IVO) as opposed to the "regular virtual orbitals" (RVO) solutions of (5). These IVO's, which therefore correspond to variationaly adjusting the orbital ϕ_l in the open-shell HF wave function for the excited state, are considerably different from the RVO's if the basis set is sufficiently flexible. In addition, the orbital ϕ_l will be different for the singlet and triplet states arising from the same orbital excitation $i \to l$. For example, the RVO π^* orbital obtained for a ground state calculation of ethylene with a *large* basis set has

$$< z^2 > \sim 225 \ (au)^2 ,$$

where z is in the π direction. The IVO π^* orbitals in the triplet and singlet $\pi\pi^*$ states, on the other hand, are characterized respectively by [68]

$$< z^2 > \sim 3 \ (au)^2 \quad \text{(triplet)}$$

$$< z^2 > \sim 27 \ (au)^2 \quad \text{(singlet)} .$$

(For a *minimum* basis set, of course, the π^* orbitals are the same, but this is an artefact of the restricted space onto which the equations have been projected).

The IVO method has three advantages. The first, which has already been pointed out, is to allow a different description of the orbitals for different excitations and different spins. The second is to improve the description of valence-type excited states when a diffuse basis set is used. In a RVO calculation the excited states may all be very diffuse even though the molecule has low-lying valence-type excited states. Finally, the IVO method is very well suited for Rydberg excited states. For such a transition the core orbitals of the upper state resemble those of the cation corresponding to the complete removal of the loosely bound electron. The excitation energy for a Rydberg state should be incorrect by roughly the same amount as in the ionization potential as given by Koopman's theorem [42]. The difference between these two quantities, *i.e.* the stabilization of the Rydberg excited state relative to ionization, should be predicted accurately. In the case of H_2O (all the excited states of H_2O are Rydberg), the experimental and theoretical stabilizations for the $^1B_1$ ($na \leftarrow 1b_1$) excited states are within 0.3 eV of each other. The calculated oscillator strengths are also found to be in good agreement with experimental values. The first $^1B_1 \leftarrow {}^1A_1$ transition has an experimental oscillator strength of 3.10^{-2} [69]. The value calculated with the IVO method is $2.86 \ 10^{-2}$ whereas the lowest energy RVO gives a value of $5.06 \ 10^{-2}$.

Essentially similar methods have been developed by Huzinaga [64,65-70] and Morokuma [71]. In Morokuma's scheme both the occupied and virtual orbitals are allowed to readjust *among themselves*; in Huzinaga's, only the vacant MO's are allowed to mix together. The Huzinaga method determines the occupied MO's by the regular HF method and then solves the modified HF-like equation which allows for the creation of a hole upon excitation. The vacant MO's are recombined within the subspace of vacant GSMO's. Such a treatment conserves the orthogonality properties, i.e. the new resultant vacant MO's remain orthogonal to the occupied ones and the excited-state wave function to the ground-state HF wave function. We will briefly develop Morokuma's method which is more general and, in addition, has been applied to formaldehyde.

Let us start with the wave function (18) for which we want to find a new set of spatial orbitals $\{\phi\}$ subject to the condition that the new MO ϕ_i is expressed as a linear combination of the occupied GSMO's ϕ_j ,and the new ϕ_l is a combination of the vacant GSMO's ϕ_p alone.

$$\phi_i = \sum_j^{occ} a_{ij}\, \phi_j \tag{20}$$

$$\phi_l = \sum_p^{vac} b_{lp}\, \phi_p \tag{21}$$

Under this restriction the ground-state wave function Ψ_0 is unchanged and Brillouin's theorem between Ψ_0 and the excited-state wave function (18) is guaranteed. The energy of the two functions (18) is given by:

$$E(i \rightarrow l) = E_0 + <l\,|\,H^{HF}\,|\,l> \; - \; <i\,|\,H^{HF}\,|\,i>$$
$$+ \, (-\,J_{il} + K_{il} \pm K_{il})\,, \tag{22}$$

where E_0 is the ground-state energy, H^{HF} the ground-state hamiltonian (6) and the upper and lower signs in (22) (and throughout this paragraph) correspond to the singlet and triplet states. The unknown coefficients a_{ij} and b_{lp} are determined by the variation method to minimize (22). This leads to the following coupled secular equations:

$$\sum_k^{occ} \{\delta_{jk}\,(H_{jj}^{HF} - \lambda_i) + B_{jk}^l\}\, a_{ik} = 0 \qquad j = 1, 2, \ldots \text{occ}; \tag{23}$$

$$\sum_q^{vac} \{\delta_{pq}\,(H_{pp}^{HF} - \lambda_l) + A_{pq}^i\}\, b_{lq} = 0 \qquad p = 1, 2, \ldots \text{vac}; \tag{24}$$

where

$$B_{jk}^l = <j\,|\,J_l - K_l \mp K_l\,|\,k> \tag{25}$$

$$A_{pq}^i = <p\,|\,-J_i + K_i \pm K_i\,|\,q>. \tag{26}$$

Eq. (23) and (24) are called the extended Hartree-Fock (EHF) equations for the orthogonalized excited state. They can be solved iteratively like the regular HF equation (5). The eigenvalues λ should be

15

called the EHF-MO energies. They include the hole—electron inter-action energy. The total energy (22) may be rewritten

$$E\,(i \to l) = E_0 + \lambda_l - \lambda_i - \{- J_{il} + K_{il} \pm K_{il}\}. \tag{27}$$

The electron—hole interaction term $\{- J_{il} + K_{il} \pm K_{il}\}$ has to be substracted due to the fact that $\lambda_l - \lambda_i$ includes this term twice. This approach can be called the electron—hole potential (EHP) method. With the additional assumption that $a_{ik} = \delta_{ik}$ the EHP method reduces itself to Huzinaga's method, which can therefore be called the hole-potential method. (By the same token, we could devise an anti-Huzinaga or electron-potential method by including only the effect of the electron potential on the occupied orbitals, *i.e.* by saying that $b_{lq} = \delta_{lq}$). This treatment has been applied to the study of hydrogen bonds in excited states of formaldehyde with water [72]. This subject will not be dealt with here, but the results on H_2CO itself are worthy of note. The vertical excitation energies are 3.106 ($^3n\pi^*$) 4.211 ($^1n\pi^*$) and 4.055 ($^3\pi\pi^*$) eV, in fair agreement with the experimental results (see Table 2). Moreover, these three excited states prefer the bent structure, which is consistent with experiment for the $n\pi^*$ states. The *relaxed* triplet $\pi\pi^*$ state, however, tends to come out lower than the *relaxed* $^3n\pi^*$ state, in disagreement with the CI results obtained with the GSMO treatments.

I.3. ESMO Methods

An alternative and potentially much simpler way of compensating for the inadequacy of the GSMO's in the description of excited states is to employ a rather limited basis set and allow separate optimization of the MO's for the excited state of interest. In other words, an independent open-shell SCF (OS.SCF) treatment, carried out for this excited state, will provide us with a set of excited-state MO's (ESMO's) that reflect the electronic characteristics of the state at hand. These ESMO's will eventually serve for CI developments. This additional freedom is partic-ularly interesting in the case of systems of low spatial symmetry where the redistribution of the electronic charge upon excitation is not restricted by symmetry requirements.

Since the electrons of spin α and β of a molecule with unbalanced spins are influenced by different systems of exchange integrals, a natural way of handling the problem is to assign different orbitals to electrons with different spins. The advantages of such a formalism have been recognized for a long time [73-75], and the first calculations were made independently by Pople and Nesbet [76] and Berthier [77-78]. In this "spin-polarized" method [79] a Slater determinant is built up from orthogonal spin orbitals ψ_i, v of which are allocated to electrons with spin α, and u to electrons with spin β. In terms of matrix elements between spin orbitals the total energy is given by:

$$E = \sum_i h_i + \frac{1}{2}(\sum_{ij} \mathscr{J}_{ij} - \sum_{ij}' \mathscr{K}_{ij}) , \qquad (28)$$

where h_i stands for monoelectronic (kinetic and nuclear attraction) integrals and $\mathscr{J}_{ij}$ and $\mathscr{K}_{ij}$ for the bielectronic (Coulomb and exchange) integrals. The prime sign on the sum over exchange integrals means that this summation has to be restricted to orbitals with the same spin factor. For clarity, let us assume that the odd (even) indices correspond to the v (u) spin orbitals with α (β) spin factor. The variationally derived OS. SCF equations are

$$(h + \overset{\text{all}}{\underset{j}{\sum}} \mathscr{J}_j - \overset{\text{odd}}{\underset{j}{\sum}} \mathscr{K}_j) \, \psi_i = \overset{\text{odd}}{\underset{j}{\sum}} \varepsilon_{ji} \, \psi_j \qquad (i = 1,3,....) \qquad (29)$$

$$(h + \overset{\text{all}}{\underset{j}{\sum}} \mathscr{J}_j - \overset{\text{even}}{\underset{j}{\sum}} \mathscr{K}_j) \, \psi_i = \overset{\text{even}}{\underset{j}{\sum}} \varepsilon_{ji} \, \psi_j \qquad (i = 2,4,....) . \qquad (30)$$

There is no need to use Lagrange multipliers between the solutions — even and odd — of each system because orthogonality is automatically ensured through spin functions. In addition, each of the two subspaces of spin orbitals can be subjected to a unitary transformation which eliminates the off-diagonal multipliers within each system. After the integration over the spin coordinates has been done, we obtain two effective eigenvalue problems connected only by the common Coulomb potential for the entire molecule:

$$\left|
\begin{array}{l}
(h + \overset{\text{all}}{\underset{j}{\sum}} J_j - \overset{\text{odd}}{\underset{j}{\sum}} K_j) \, \phi_i = \varepsilon_i \phi_i \\[2em]
(h + \overset{\text{all}}{\underset{j}{\sum}} J_j - \overset{\text{even}}{\underset{j}{\sum}} K_j) \, \phi_i = \varepsilon_i \phi_i .
\end{array}
\right. \qquad (31/32)$$

At this stage the HF method is "unrestricted" since no peculiar relation exists between the spin orbitals except for their orthogonality. However, constraints may be introduced (the HF scheme becoming therefore of a "restricted" type); these are of two basic types [80]:

a) *spin-equivalence restriction:* if we want to assign two electrons with opposite spins to the same one-particle energy level ε_i, we must assume that each spatial orbital ϕ_i can be combined with either the α or β spin functions to form two degenerate spin orbitals ψ_i and ψ_i'.

b) *symmetry-equivalence restriction:* if, similarly, we want to force $2p$ electrons to occupy the p spatial orbitals of a p-fold degenerate level ε_i, we must properly connect these p spatial orbitals. For example, the two MO's belonging to the Π representation of a linear molecule differ one from another in their angular factors, $e^{i\varphi}$ and $e^{-i\varphi}$.

Let us examine the consequences of the spin restriction [79]. We must identify the spatial part of the rth solution of Eq. (29) with the spatial part of, say, the $(r+1)$th solution of (30). After integrating over the spin coordinates, we obtain a "restricted" system of two coupled equations by replacing u Eq. of (29) and (30) by their sum and retaining the $(v-u)$ last Eq. of (29).

$$\{h + \sum_{l=1}^{u}(2J_l - K_l) + \sum_{n=u+1}^{v}(J_n - \frac{1}{2}K_n)\}\phi_k = \sum_{l=1}^{u}\varepsilon_{lk}\phi_l + \frac{1}{2}\sum_{n=u+1}^{v}\varepsilon_{nk}\phi_n.$$

$$k = 1,2,\ldots u. \tag{33}$$

$$\{h + \sum_{l=1}^{u}(2J_l - K_l) + \sum_{n=u+1}^{v}(J_n - K_n)\}\phi_m = \sum_{l=1}^{u}\varepsilon_{lm}\phi_l + \sum_{n=u+1}^{v}\varepsilon_{nm}\phi_n.$$

$$m = u + 1,\ldots v \tag{34}$$

Here again, it is possible to eliminate, by two suitable unitary transformations, the Lagrange multipliers connecting the doubly occupied (or the singly occupied) MO's among themselves. However, it is not possible to annihilate those interconnecting the two sets without destroying the orthogonality of the solutions of (33) and (34). The final result is a system of two coupled HF equations, the first involving a core hamiltonian F^C, the second an open-shell hamiltonian F^O:

$$F^C \phi_k = \varepsilon_k\phi_k + \frac{1}{2}\sum_{n=u+1}^{v}\varepsilon_{nk}\phi_n. \tag{35}$$

$$F^O \phi_m = \varepsilon_m\phi_m + \sum_{l=1}^{u}\varepsilon_{lm}\phi_l. \tag{36}$$

Two general approaches have been devised to solve this system. For a large class of excited states, the first — Roothaan's method [81-82] — rigorously reduces Eq. (31) and (32) to a unique eigenvalue problem by absorbing the off-diagonal Lagrange multipliers into a new effective hamiltonian. The second — Nesbet's method [83] — defines a suitable but

18

approximate effective hamiltonian whose eigenfunctions will be the α spinorbitals. The β electrons will then be forced to occupy the same set of spatial orbitals. These two methods will now be briefly discussed.

I.3.1. Roothaan's Treatment

The total wave function of the excited state is written as a linear combination of Slater determinants. Each of these consists of two parts: a closed shell $\{\phi_C\}$ of doubly occupied orbitals $(k, l..)$, and an open–shell of singly occupied orbitals $(m,n..)$ chosen among a subspace $\{\phi_O\}$. The set $\{\phi_C, \phi_O\}$ is orthogonal. Let us suppose that the energy of this wave function may be written as:

$$
\begin{aligned}
E = 2 \sum_k h_k &+ \sum_{kl} (2 J_{kl} - K_{kl}) \\
&+ f \{2 \sum_m h_m + f \sum_{mn} (2 a J_{mn} - b K_{mn}) \\
&+ 2 f \sum_{km} (2 J_{km} - K_{km}) .
\end{aligned}
\tag{37}
$$

(37) is clearly the sum of three terms: the energies of the separated closed and open shells and their interaction energy; f, a, b are specific numbers depending on the distribution of unpaired electrons. In particular, f, the ratio of the number of occupied and available open-shell spin orbitals, is the fractional occupation number of the $\{\phi_O\}$ subspace. Two new operators the "Coulomb- and exchange-coupling operators" L_i and M_i, are defined in the following manner:

$$
L_i \phi = <\phi_i | f \sum_m J_m | \phi > \phi_i + <\phi_i | \phi > \cdot f \cdot \sum_m J_m \phi_i
\tag{38}
$$

$$
M_i \phi = <\phi_i | f \sum_m K_m | \phi > \phi_i + <\phi_i | \phi > \cdot f \cdot \sum_m K_m \phi_i .
\tag{39}
$$

Finally, for each of the four types of operator $A = J,K,L,M$ we introduce the total closed-shell operator A_C, the total open-shell operator A_O and their sum A_T:

$$
A_C = \sum_k A_k, \quad A_O = \sum_m A_m, \quad A_T = A_C + A_O
\tag{40}
$$

The great merit of this formalism is to enable us to express the Lagrange multiplier ε_{km} connecting the two subspaces in a simple matrix form:

$$
\varepsilon_{km} = - f \cdot <\phi_k | 2 \alpha J_O - \beta K_O | \phi_m >
\tag{41}
$$

19

where α and β are simply numerical coefficients

$$\alpha = \frac{1-a}{1-f} \qquad \beta = \frac{1-b}{1-f} \qquad (42)$$

Consequently, Eq. (35) and (36) collapse into a unique eigenvalue problem:

$$\{h + (2\,J_{\mathrm{T}} - K_{\mathrm{T}}) + 2\,\alpha(L_{\mathrm{T}} - J_{\mathrm{O}}) - \beta(M_{\mathrm{T}} - K_{\mathrm{O}})\}\,\phi_i = \varepsilon_i\,\phi_i, \qquad (43)$$

which gives the closed-shell as well as the open-shell wave functions. The energy (37) may be rewritten:

$$E = \sum_k (h_k + \varepsilon_k) + f \sum_m (\varepsilon_m + h_m)$$

$$- f \sum_{km} (2\,\alpha\,J_{km} - \beta\,K_{km}) - f^3 \cdot \sum_{mn} (2\,\alpha\,J_{mn} - \beta\,K_{mn}). \qquad (44)$$

The calculation of the lowest singly excited triplet (of a given symmetry) of a molecule possessing a closed-shell ground state is possible according to Roothaan's scheme. For singlet states, however, the initial and final MO's of the single excitation process must transform according to *different* irreducible representations (at least one of which is non-degenerate) of the symmetry group of the molecule. (This condition goes beyond the present scheme and, more generally, is related to the basic difficulty of maintaining orthogonality between two SCF functions of the same symmetry.)

In the case of a planar (C_{2v}) formaldehyde molecule, direct open-shell calculations are then possible for the $^3A_1(\pi\pi^*)$, $^{1,3}A_2(n\pi^*)$, $^{1,3}B_1$ ($\sigma\pi^*$) and $^{1,3}B_2(n\sigma^*)$ excited states. The important $^1A_1(\pi\pi^*)$ does not satisfy the necessary condition since the π and π^* orbitals both belong to the B_1 irreducible representation. Even though singlet-state calculations are only developed here for comparison with the triplet, it is worthwhile mentioning that, to circumvent this problem, Basch et al. [85] have suggested using the triplet A_1 ESMO's to describe the parent singlet. Because of the reduced symmetry of the bent formaldehyde structure, OS.SCF calculations can be done only for the three $^3A'(\pi\pi^*)$ and $^{1,3}A''(n\pi^*)$ excited states [53]. As may be seen in Table 4, this procedure gives the same ordering as the CI(GSMO) method. However, the total energy of the excited states is, as expected, more effectively lowered than by the CI(GSMO) method. The vertical transition energies of the $n\pi^*$ states are even smaller than the experimental ones. This is due to the

Table 4. Results of the OS.SCF and CI(ESMO) treatments of the $^{1,3}n\pi^*$ and $^3\pi\pi^*$ excited states of formaldehyde. They are taken from the work of Buenker and Peyerimhoff [53] and Fink [90]

Property		OS.SCF single configuration treatment (Roothaan) B.P.	CI(ESMO) treatment (Roothaan) B.P.	OS.SCF single config-uration (Nesbet) F.	CI(ESMO) treatment (Nesbet) F.
Out-of-plane bending angle θ (degrees)	$^3A_2$	32.0°	32°		
	$^1A_2$	31.1°	30°		
	$^3A_1$	20°	20°		
Vertical excitation energies (eV)	$^3A_2$	2.24 2.67[1]	3.01 3.23[1]	2.58	2.41
	$^1A_2$	2.63	3.43	3.23	2.96
	$^3A_1$	4.20	4.99	—	—

[1] The corresponding results of Whitten and Hackmeyer [30] have been added when available.

fact that the correlation energy for an open-shell state is undoubtedly smaller than that of the ground state. On the other hand, the calculations do succeed in obtaining potential energy minima for the $n\pi^*$ states in the bent conformation. The bending angles of 32.0° ($^3A''$) and 31.1° ($^1A''$) are in good agreement with the experimental ones.

At this point it is clear that the CI(GSMO) procedure invariably overestimates the vertical excitation energies (because of the inadequacy of the GSMO's for the description of excited states) and that the OS.SCF underestimates them (due to the unequal correlation energies of the ground and excited states). A CI(ESMO) treatment which combines these two limiting procedures would give vertical excitation energies which fall between these two extremes. A limited CI on each state of interest based on its own set of MO's will give a better description of the differences in charge distributions than when a unique set of MO's is used for all excited species, will correct the biased treatment of the correlation energies and will ensure automatic orthogonality between states of the same symmetry and multiplicity.

In their CI(ESMO) treatment of the excited states of formaldehyde [53], Buenker and Peyerimhoff use the same partitioning technique [86] of the MO's which was chosen at the CI(GSMO) level, but include *all* triple and quadruple excitations. The number of configurations included in the CI is then of the order of two hundred. Their results are summarized in Table 4. The combined treatment gives by far the best agreement with experimental results. The discrepancies in the vertical transition energies are only 0.11 and 0.07 eV for the singlet and triplet $n\pi^*$ states, respectively. These two states are definitely found to be bent, the triplet state being slightly more bent than the singlet, in good accord with our experimental knowledge, even though the calculated difference between the two bending angles appears to be much smaller than the measured one. Good agreement with experiment is also obtained for the inversion barriers in the $n\pi^*$ states. The calculated singlet and triplet barriers are 0.090 and 0.073 eV, the experimental ones 0.096 and 0.044 eV [35]. Here again the CI(ESMO) does not find as sharp a difference as that inferred experimentally. Finally, the equilibrium CO bond lengths are approximately 0.10 au greater than the experimental values. This error is in the same direction and of the same order as that found for the ground state and may be traceable to the tendency of the Gaussian basis function to overestimate these quantities [88].

I.3.2. Nesbet's Formalism

Instead of solving two systems of simultaneous equations, a sufficient approximation might be to calculate the spatial orbitals ϕ assigned to the electrons with a given spin, say α, from a suitable effective hamiltonian and then to force the electrons with spin β to occupy the same set of spatial orbitals. The total wave function is then built up from ortho-normal doubly and singly occupied orbitals and can easily satisfy all spin and symmetry restrictions. Of course, the expression taken as the effective hamiltonian is only an approximation and does not correspond to the actual form of the assumed wave function. A major consequence is that the total energy cannot reach its absolute value.

The general form of the effective hamiltonian H_e includes an arbitrary factor x multiplying those exchange integrals which involve the two odd electrons:

$$H_e = h + \sum_{j}^{occ} (2 J_j - K_j) + (J_i - xK_i) + (J_l - xK_l) . \quad (45)$$

If the operator H_e were to act separately on each electron in a doubly occupied orbital j, the optimum value of x would be either 1 or 0, depending on the spin of the electron considered. It would therefore be reasonable to choose the average value $x = 1/2$ and write:

$$H_e = h + \sum_{j}^{occ} (2 J_j - K_j) + (J_i - \tfrac{1}{2} K_i) + (J_l - \tfrac{1}{2} K_l) . \quad (46)$$

Alternatively x may be considered as an additional parameter to be optimized. For example, in their study of the geometrical isomerization of cyclopropane, Salem and his collaborators have used an optimized value $x = 0.42$ for both the lowest singlet and triplet states of the diradical intermediates [89].

If we use the general form (45) of the effective hamiltonian, the energy of the excited $i \to l$ triplet state is given by:

$$E(^3i \to l) = \sum_{j}^{occ} (\varepsilon_j + h_j) + \tfrac{1}{2} (\varepsilon_i + h_i + \varepsilon_l + h_l)$$
$$+ (x - \tfrac{1}{2}) \sum_{j}^{occ} (K_{ji} + K_{jl}) + (x - 1) K_{il} \qquad (47)$$
$$+ \tfrac{1}{2} (x - 1) (J_{ii} + J_{ll}) .$$

(The energy of the parent singlet is obtained by simply adding $2 K_{il}$).

The advantage of this method is its simplicity and low cost. For that reason, it is very often used to minimize excited-state structures or to scan large regions of excited-state potential energy surfaces. The major part of the photochemical problems examined in the second section of this article is treated within this framework. We shall, for example, examine the dissociation of formaldehyde into radical or molecular products. To treat this problem Fink [90] chose to carry out a Nesbet OS.SCF calculation for the $n\pi^*$ triplet. The resulting MO's were then used (a) to describe the parent singlet state and (b) as an expansion set for the CI treatments. The following excitations were included: all excitations from the singly occupied $n(2\,b_2)$ and $\pi^*(2\,b_1)$ MO's into the three lowest vacant orbitals $(6\,a_1, 3\,b_2, 3\,b_1)$, the single excitations from the two upper closed-shell orbitals $(5\,a_1, 1\,b_1)$ to $2\,b_2$ and $2\,b_1$, and finally excitations from one singly occupied orbital to the other. These various excitations give a total of 50 determinants in the CI. The results are given in Table 4. They are not very good, but Fink had to greatly reduce the accuracy of the calculations in the interest of being able to trace out portions of the potential energy surfaces.

An advantage of the single effective hamiltonian is that its form may be adapted to particular situations. To illustrate this point let us consider briefly the crossing of two singlet states of different symmetry, the first one being a closed-shell singlet (i^2) (l^0), the second an open-shell singlet (i^1) (l^1). The HF operator of the closed-shell system

$$H^{HF} = h + \sum_{j}^{occ} (2 J_j - K_j) + (2 J_i - K_i) \qquad (6)$$

provides us with a good representation of the occupied orbitals of the closed-shell but a poor representation of those of the open-shell. The energy of the open-shell structure will then be too high and the crossing is expected to occur too far along the crossing coordinate. In Nesbet's formalism the HF operator for the open-shell system,

$$H_e = h + \sum_j^{occ} (2 J_j - K_j) + (J_i - \tfrac{1}{2} K_i) + (J_l - \tfrac{1}{2} K_l) , \qquad (46)$$

provides a fairly good representation of the open-shell orbitals but is inadequate for the ground state, which will be therefore too high in energy. The crossing now occurs too soon along the crossing coordinate. A simple solution of this dilemma is to use an HF hamiltonian *intermediate* between (6) and (46) [91]

$$H_i = h + \sum_j^{occ} (2 J_j - K_j) + \tfrac{3}{2} (J_i - \tfrac{1}{2} K_i) + \tfrac{1}{2} (J_l - \tfrac{1}{2} K_l) . \qquad (47)$$

This hamiltonian treats both states with equal bias. The crossing point, which depends on the *relative* energies of both states, will be correctly determined (even if its absolute energy is still too high).

At this stage of our study it is clear that the best procedure to describe an excited state involves two steps:

1. A closed-shell SCF calculation of the one-determinantal ground state wave function ψ_0 (energy E_0), which will give a set of occupied and vacant GSMO's. These, or some of these, will then be used to generate configurations which will mix with the ground-state wave function in a CI treatment. The result of this treatment is a ground-state wave function expressed as a linear combination of configurations (the leading term being ψ_0) whose energy lies below E_0.

2. An open-shell SCF calculation of the low-lying state of interest will give a set of ESMO's adapted to this excited state. These ESMO's will subsequently be used to perform a CI treatment again involving several configurations of the same symmetry as the state at hand. For triplet states, Roothaan's rigorous OS.SCF treatment should be used, though Nesbet's approximate hamiltonian is less time-consuming when potential energy surfaces have to be explored.

The CI treatment finds its justification in the intimate nature of the Hartree-Fock methodology [92,93]. An electron may be pictured as surrounded by a "Coulomb" hole outside which interelectronic interaction repels all the other electrons, of whatever spin. These two-electron inter-actions influence the wave function through the Coulomb and exchange integrals. Indeed, the repulsive potential becomes singular as the relative distance between the two electrons vanishes and consequently modifies the wave function near these singularities and introduces a *correlation* between each pair of electrons. In terms of probabilities, this correlation means that the *joint* probability of finding electron 1 in volume element $d\tau_1$ and electron 2 in $d\tau_2$ is not equal to the product of the separate probabilities of finding electron 1 in $d\tau_1$ and electron 2 in $d\tau_2$. Now, this equality (stochastic independence of probabilities) is required by the structure of a determinantal wave function. Such a HF wave function therefore does not properly describe the correlated motion of the elec-trons.

Another effect, however, intervenes. The many-electron wave func-tion (such as (1)), being a determinant, is totally antisymmetric with respect to the exchange of two electrons and therefore obeys Pauli's principle. This means that (1) vanishes when two electrons having the same spin also happen to have the same spatial coordinates. An elec-tron may then be pictured as surrounded by another hole (Fermi hole) entry to which is forbidden to electrons of the same spin.

We see that the Fermi hole for electrons of the *same* spin is an approx-imate substitute for the Coulomb hole and that the correlation problem for electrons of the same spin is, at least partially, eliminated. The situa-tion is far more dramatic for electrons of different spin — and two elec-trons paired in the same molecular orbital enter this category — for which there is no Fermi hole. Two such electrons have a definite probability of being in the same volume element and the correlation problem here is particularly acute.

The total correlation energy may be approximated by a sum over pair correlation energies [94]: (independent electron-pair approximation: IEPA, note that the ε's here are not Lagrange multipliers).

$$E_{corr} \sim \sum_i \varepsilon(i,i) + \sum_{i<j} \{{}^1\varepsilon(i,j) + {}^3\varepsilon(i,j)\} . \tag{48}$$

$\varepsilon(i,i)$ is the intrapair correlation energy for the two electrons in the *i*th MO. $\varepsilon(i,j)$ is the interpair correlation energy (for triplet or singlet coupling, depending on the spins of the interacting electrons in the MO's *i* and *j*, $i \neq j$).

The intrapair correlation energy is usually greater than the interpair correlation energy. For this reason the correlation energy of an open-shell system is smaller than that of the closed-shell ground state, since the excitation process destroys an intrapair correlation energy term.

One of the methods [93] for calculating correlation energies is the configuration interaction treatment. Accurate vertical transition energies may be obtained only if the ground and excited states are treated at the same level. If both states are described at the single configuration level, the smaller open-shell correlation energy leads to an underestimation of the transition energy. A more satisfactory solution is to carry out an appropriate CI treatment on both states, based on the proper sets of MO's, so that the correlation energies are, at least in part, included in the total energies. As we pointed out previously, the best results are obtained from a CI (GSMO) treatment of the ground state and a CI (ESMO) treatment of the excited state.

II. Potential Energy Surfaces of Triplet States

The study of photochemical processes can be conducted at three different levels. The *first* may be called the *molecular orbital (MO) level* [45]. The use of simple MO arguments allows us to estimate the nuclear geometries at which the potential energy surfaces of two different states undergo (generally avoided) crossings. In such regions "jumps" from one surface to the other (intersystem crossing and internal conversion) should be particularly easy [96,98]. Similar arguments can be used to estimate the position of potential energy barriers in the surface of a given excited state and therefore to distinguish between "allowed" and "forbidden" processes [99,103]. Finally, even along reaction paths which at first appear "allowed", the occurrence of "abnormal" orbital crossings may result in the appearance of additional barriers [104,106]. At this first level of description, minima and barriers appear as a consequence of the nodal properties of MO's and, as such, are said to be "imposed by symmetry". There are undoubtedly other barriers whose existence would be revealed by accurate calculations. These quantitative arguments are supported by Hückel-type calculations, which unfortunately do not distinguish properly between states of different multiplicities [97,98,99,107]. Addition of the interelectronic repulsion terms destroys the simplicity of the Hückel scheme but permits differentiation between singlets and triplets. As a consequence most of the aforementioned crossings are avoided. It is important to recall here that the singlet and triplet states resulting from the same spatial excitation generally do not have the same electronic characteristics and therefore react differently [108,109]. Finally, purely perturbational treatments which basically estimate the shape and the slope of certain regions of potential energy surfaces [107,110,113] may also support MO analysis.

The *second* stage of the study of photochemical reactions, which is the subject of this section, emphasizes the importance of *electronic states*. This level of description, based on sufficiently accurate "ab-initio" calculations, can deal with two types of problems. The first problem is the determination of the potential energy surfaces of the low-lying excited states of a single molecule and it provides us with two families of results:

a) the equilibrium geometries of each of the various triplets of a molecule. This question has already been the subject of numerous calculations on diatomic (H_2 [114,117], He_2 [118], O_2 [119,120], Li_2 [121], LiH [122,123], NaLi [124], SiO [125], MgO [126]) and polyatomic molecules (H_4 [127], O_3 [128,130], cyclopropane and trimethylene [89,131], nitrenes [132], C_2H_2, HCN, FCN, H_2CO, F_2CO, HCF, HNO, FNO [133]).

b) the behavior of the triplet states of a molecule when a dissociation process occurs (cyclopropane [89], ketene [134]).

The second type of problem involves the study of the potential energy surfaces of a reacting system, *i.e.* a "supermolecule" involving two molecules, one of which has been primarily excited. This classification will be the skeleton of the following section.

Before developing it, we ought to mention that these two stages — MO's and electronic states — are "static" levels of description. They should be followed by a *third* level which is the dynamical analysis of the behavior of the molecule(s) along the calculated potential energy surfaces. The static surfaces can be expected to predict the nature of many or most possible products. The relative quantum yields, the relative rates, and the possible wavelength dependence of the formation of these products can only be obtained from a dynamical study involving, in particular, a detailed analysis of the "radiationless" jumps between various surfaces.

II.1. Potential Energy Surfaces for the Lowest $^3\pi\pi^*$ State of Polyenes

The triplet state of ethylene is the simplest organic triplet. Its π system results from the interaction of two $p\pi$ orbitals, ϕ_1 on carbon C_1 and ϕ_2 on carbon C_2, each of them containing a single electron of spin α. Two limiting cases may be imagined: in the first the axes of the two atomic orbitals ϕ_1 and ϕ_2 are perpendicular. There is no overlap, and therefore no combination between ϕ_1 and ϕ_2. This the the *orthogonal* form of the triplet. In the second case the axes are parallel and the two atomic orbitals ϕ_1 and ϕ_2 combine (conjugate) to form two molecular orbitals, each singly occupied by an α electron. This is the *planar* form of the triplet. Simple first-order perturbation theory [135,137] shows that the conjugation process between ϕ_1 and ϕ_2 is destructive, since it in-

27

volves a loss in π energy. This loss is mainly due to the fact that the anti-bonding combination $\phi_1 - \phi_2$ is more destabilized than the bonding combination $\phi_1 + \phi_2$ is stabilized, and its magnitude is proportional to the square of the overlap integral $S_{12} = <\phi_1|\phi_2>$. Consequently, any geometrical change which reduces the magnitude of S_{12} stabilizes the planar triplet state. Thus the C_1-C_2 bond length in the planar triplet state of ethylene should be longer than the length of a single bond and relatively long bonds should be present in the triplet states of conjugated polyenes. A more efficient method of reducing S_{12} is to twist one of the p_π orbitals through 90° around the C_1C_2 axis. The (unconjugated) orthogonal triplet thus obtained is more stable than the (conjugated) planar triplet.

These straightforward predictions have been verified by "ab-initio" calculations. Kaldor and Shavitt [138] have used the Roothaan OS.SCF treatment (without CI since there is no other low-lying triplet of appropriate symmetry) to obtain a vertical excitation energy for the $^3B_{3u}$ triplet (T) of planar (D_{2h}) ethylene of 4.50 eV. As expected, this value is slightly lower than the experimental determination (4.60 eV) [139]. (Other typical calculated values are 4.32 eV [140] and 4.19 eV [141]). The twisting potential energy curve (the CC bond length being kept constant) of the $^3B_3$ triplet state is represented in Fig. 1 together with the ^1A ground state (N) curve (closed-shell treatment with CI). The energies of the orthogonal (D_{2d}) ground state $(^1B_1)$ and triplet state $(^3A_2)$ are respectively 3.61 and 3.26 eV above the planar ground state.

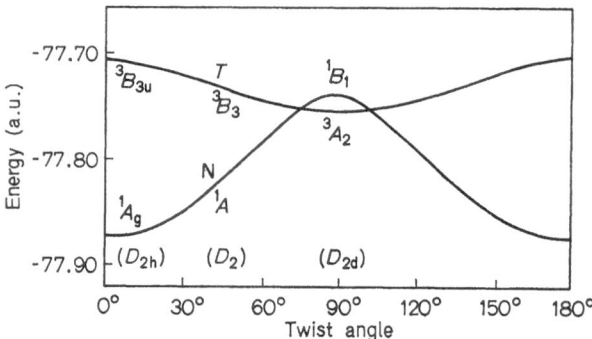

Fig. 1. Twisting potentials of the ground state (N) and the $\pi\pi^*$ triplet state (T) of ethylene [138]

As predicted, the orthogonal triplet is more stable than the planar triplet (by 1.34 eV) and even becomes the ground state of the orthogonal structure. How do these results compare with experiments? Two routes have been proposed for the geometrical isomerization of ethylene.

1. The first route follows the ground-state N curve all the way. The energy of the $N(^1B_1)$ orthogonal structure then corresponds to the activation energy of the *cis–trans* isomerization process. A value of 2.67 eV has been reported for the isomerization of dideuterio-ethylene [142].

2. An alternative route has been suggested involving radiationless transitions between the N and T states at their intersecting points [143]. No reliable estimate could be found of the barrier to rotation by this route. The estimate of Eyring and coworkers [143,144] (1.2 eV) is obtained for relatively complex molecules (styrene) and may be related to rather complicated mechanisms. The value of 0.9 eV [136,145] was found as the activation energy of *cis–trans* isomerization of 2-butene but is very inconclusive as to the order and the mechanism of the reaction.

It is readily seen from the theoretical calculations that the barrier in the T state is of the order of 3.4 eV, much higher than that inferred experimentally. The discrepancy may come from the fact that more complicated systems than ethylene were investigated, though it is difficult to understand why ethylene and 2-butene would follow completely different isomerization paths. The second possibility is a substantial stabilization of the orthogonal triplet by further distortions [146]. Such a stabilizing deformation might well be the flapping of the CH_2 groups that corresponds to a pyramidalization of the C atoms [147,149]. The "ab-initio" study of these distortions has been recently done by Baird and Swenson [150] using the OS.SCF method of Roothaan. When a *minimal* basis set $\{g_m\}$ of Gaussian functions is used, the planar triplet is found to have an equilibrium bond length of 1.55 Å, greater than the single-bond length, as predicted by the perturbation scheme. The planar structure is further stabilized by a *cis*-flapping motion 1 of 35° (stabilization energy of 5.1 kcal/mole) or a *trans*-flapping motion 2 of 33° (3.6 kcal/mole). Here again, these motions decrease S_{12} and the destructive conjugation. The orthogonal triplet has a shorter equilibrium bond length of 1.48 Å, due to the weaker repulsive interaction between the two unpaired electrons, and presents a very shallow minimum (0.8 kcal/mole) for a flapping 3 of 23°. If we now use an *extended* basis set, the planar triplet again presents local minima for a *cis*-flapping of 26° (1.3 kcal/mole)

1   2   3

or a *trans*-flapping of 25° (0.7 kcal/mole) but no such stabilization is found for the orthogonal triplet. On the contrary, the orthogonal conformation with a flap angle of 25° is 1.6 kcal/mole less stable than the undistorted orthogonal form. The contrast between the predictions of the two basis sets parallels that encountered in the calculation of the inversion barrier of NH_3 [15], where the energetic preference for a pyramidal rather than a planar structure is overestimated at the minimal basis set level and underestimated at the extended basis level. If this analogy is appropriate, we may expect that the use of more elaborate basis sets (including *d*-type polarization functions) will give a pyramidal conformation (1 and 2) for the planar triplet (with a corresponding stabilization between 1 and 5 kcal/mole) and possibly a flapping of the orthogonal triplet, which would bring a small extra stabilization of the order of 1 kcal/mole. In any case, these values cannot reconcile the calculated and experimental values for the isomerization barrier of ethylene.

If we now turn our attention toward the higher members of the polyenic family, we find that *trans*-butadiene still qualitatively follows

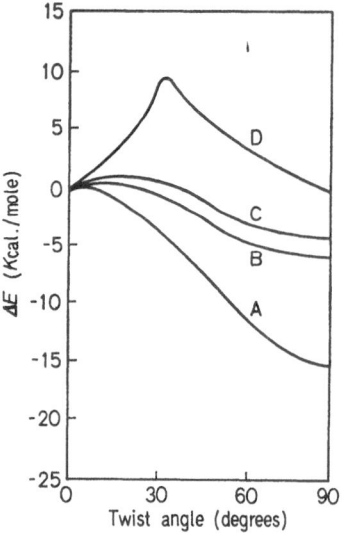

Fig. 2. Potential energy curves for C=C bond twisting in polyene triplets: ethylene (*A*), 3.4 bond and 1.2 bond in *trans—trans* hexatriene (*B* and *D*, respectively), *trans* butadiene (*C*) [137]

the behavior predicted by the simple perturbation scheme (Fig. 2), but also that longer polyenes present several new features. First, the preference for a triplet twisted around an *internal* double bond is much greater than expected and is much more favorable than the twist around a *terminal* double bond, which is much smaller than expected. Second, we note (Fig. 2) the appearance of intermediate barriers (peaking at ~30°) to rotation around a terminal bond. These effects have been clearly explained by Baird and West [137] who considered the bonding character in a system such as hexatriene. The unbranched hexatriene molecule may be viewed as two allylic systems destructively conjugated accross the 34 bond. The rotation around this rather "single" bond in the $\pi\pi^*$ triplet state does not require a drastic reorganization of the electronic

system, but does relieve the destructive conjugation. No barrier forbids this stabilizing twist and, as in ethylene, the 90° form is the most stable. The twisting around the 12 terminal bond, however, involves a complete reorganization of the electronic system toward a "methyl-pentadienyl" structure and then leads to an intermediate barrier. The reduction in destructive conjugation brought about by the twist (5 kcal/mole) is cancelled by the smaller inherent stability (4—5 kcal/mole) of the methyl-pentadienyl system relative to the allyl—allyl system. The planar and the 90° (1—2) twisted forms of the $\pi\pi^*$triplet state of hexatriene are therefore nearly degenerate.

II.2. Morphology of the $^3n\pi^*$ and $^3\pi\pi^*$States of Acrolein

The $\alpha\beta$ unsaturated ketones present two types of low-lying states: the $\pi\pi^*$ triplet state, which closely corresponds to the $\pi\pi^*$ triplet of polyenes, and the $n\pi^*$ triplet state, which results from the promotion of an electron out of the doubly occupied non-bonding lone-pair orbital localized, at the simplest level of description, on the oxygen atom. Let us first consider the behavior of these two triplets during the twist around the ethylenic double bond of the acrolein molecule, the first member of the enone family. If the terminal CH_2 group is twisted by 90°, two non-bonding MO's appear (which correspond to ϕ_1 and ϕ_2 in our description of ethylene): the first (namely y, A' symmetry) is a nearly pure $2p_y$ atomic orbital located on C_1, the second (namely π', A'' symmetry) is an allyl-type orbital with a nodal plane passing near the C_3 carbon atom (see Fig. 3). The potential energy surfaces have been calculated by using the effective single hamiltonian of Nesbet's method and a limited CI treatment whenever two triplet states of the same symmetry are

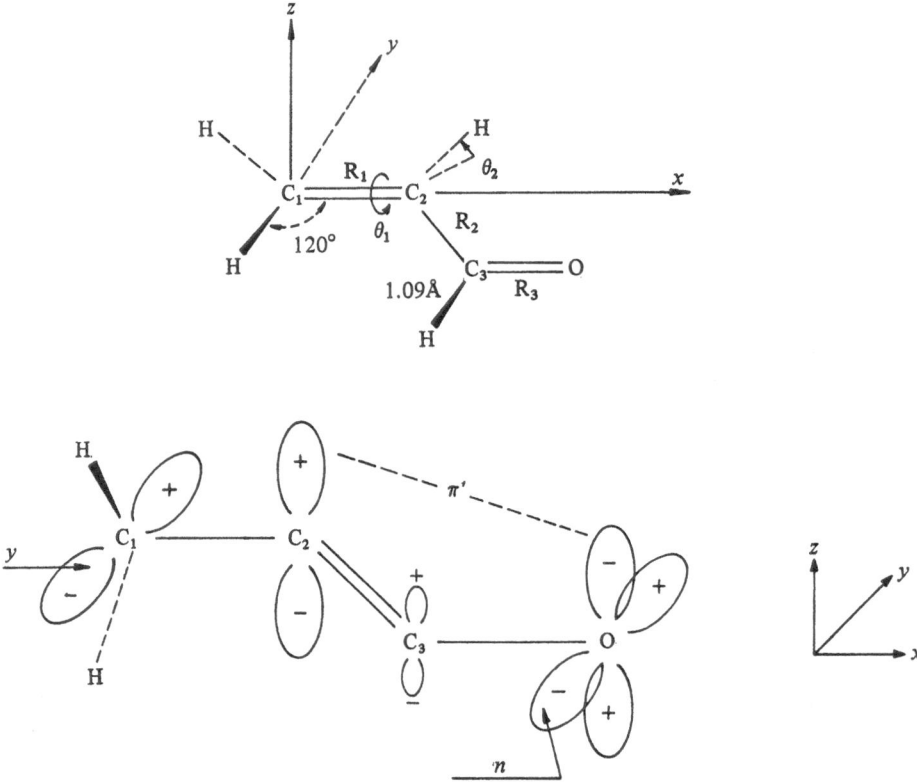

Fig. 3. "Frontier" orbitals in the 90° twisted form of acrolein [153]

close in energy [152,153]. During the rotation the $\pi\pi^*$ triplet is stabilized (Fig. 4) by 24.5 kcal/mole and apparently correlates with the $\pi'y$ triplet state of the orthogonal form. As in the case of ethylene, this orthogonal triplet is the ground state of the twisted structure. (The $\pi'y$ triplet lies 1.9 kcal/mole below the parent singlet state). The $n\pi^*$ triplet on the other hand is destabilized (by 26.4 kcal/mole) in this process and correlates with the ny triplet of the twisted acrolein. For all intermediate angles of twist the two triplets have the same symmetry (no symmetry at all) and therefore must be mixed in the unavoidable CI treatment. The exact correlation diagram thus links $^3n\pi^*$ to ^3ny and $^3\pi\pi^*$ to $^3\pi'y$ when the vertically excited $\pi\pi^*$ triplet is above the $n\pi^*$ triplet. This is both calculated and confirmed experimentally [201] to be the case, the vertical excitation energies of the $\pi\pi^*$ and $n\pi^*$ triplet states of acrolein being 70.4 and 69.6 kcal/mole, respectively.

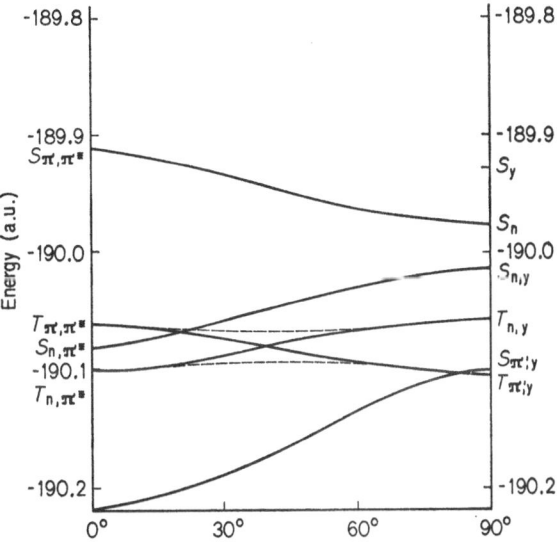

Fig. 4. Theoretical twisting potentials for the ground and lowest excited states of s-*trans* acrolein [153]

The potential energy curves for the $\pi\pi^*$ triplet of acrolein and, more generally, of higher enonic systems are qualitatively similar to those of the corresponding polyenes [154]. The same rationalization, based on the idea of destructive conjugation, also holds. The minor differences between the two parent systems arise mainly from the greater strength of the C=O π bond (78.3 kcal/mole) relative to the C=C π bond (58.1 kcal/mole). A first consequence of this is the near localization of the excitation in the C=C bonds in long polyenones such as $CH_2(CH)_{10}O$, a useful model for the retinal molecule [155]. For this model system the energy difference between the 0° and 90° twisted forms for the lowest $^3\pi\pi^*$ state has been calculated semi-empirically [19,154]. As shown below, all twisted conformations are predicted to be less stable than the planar structure but rotations around the 9.10 and 11.12 C=C bonds require only a very small energy (0.4 kcal/mole)

A second consequence is that the acroleinic structures (a) and (b)

$$\dot{C} \text{—} \dot{C}\text{—}C\text{=}O$$

(a)

$$\dot{C} \text{—} C\text{=}C\text{—}\dot{O}$$

(b)

33

are no longer in close resonance as they are in butadiene. Instead (a) is much more stable than (b). Indeed, it has been shown from bond dissociation energy studies that the resonance stabilization in free radicals like $R_1C(=O)\dot{C}HR_2$ is much smaller than that for allylic radicals [156]. Another property of the $^3\pi\pi^*$ state is worthy of note: when the enone group is incorporated into a ring (cyclohexenone, steroidal enones ...), the ring strain partly (if the enone belongs to the A ring of steroids) or almost completely (B ring of steroids) prevents the CH_2 terminal group from twisting. In order to relieve the destructive conjugation trapped in the planar structure, the molecule must seek more subtle ways, one of them being the out-of-plane motion of the hydrogen atom attached to atom $C_2(H_3)$. To be efficient, this motion must, of course, take place in

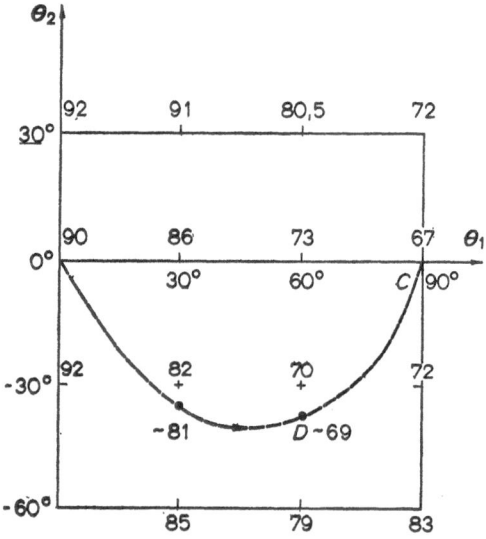

Fig. 5. Influence of the out-of-plane motion of the H_3 hydrogen atom in the case of the $\pi\pi^*$ triplet state of acrolein [153]

the direction opposite to the CH_2 twist. Fig. 5 shows the effect of this distortion [153]. The most stable point of the surface still corresponds to θ_1 (twist motion) $= 90°$ and θ_2 (out-of-plane motion) $= 0°$, but the best stabilizing path reaching this point C involves an out-of-plane motion of H_3 in the direction opposite to the twist ($\theta_1 > 0$, $\theta_2 < 0$). If then the strain in a molecule such as, say, cyclohexenone allows a twist θ_1 no greater than $60°$, the minima of the $^3\pi\pi^*$ state will be located at D, a conformation in which the out-of-plane motion ($\theta_2 = -40°$) brings an added stabilization of 4 kcal/mole. At this point the originally pure $2p_z$ atomic orbitals on C_1 and C_2 have acquired some s character. The axes of these two new hybrids are almost perpendicular; such a triplet presents a more or less orthogonal conformation.

In sharp contrast to the $\pi\pi^*$ triplet state, the $n\pi^*$ *triplet* of acrolein is destabilized by the twisting motion. The π bonding in this state, in acrolein or polyenones, is similar to that in the ground state of the corresponding free radicals $R_1R_2\dot{C}H$, in which the hydrogen atom is substituted — and only weakly perturbed — by an alcohol or ether group. This means that the predominant resonance structures in this system are (c) and (d). The

$$C=C-\dot{C}-\dot{O}\dot{C}-C=C-\dot{O}$$

(c) (d)

hypothesis that delocalization of the oxygen π lone pair is unimportant energetically in radicals is supported by the C—H bond dissociation energies for alcohols of the type R_1R_2CHOH (R_1,R_2 = alkyl or H), which are found to be almost identical to those of $R_1R_2CHCH_3$ compounds [157]. The OH group is not much more effective in stabilizing a π radical than is a methyl group. The first consequence is that the rotation of the terminal CH_2 group destroys the constructive stabilization of the allylic group. As a result, the "free-radical ether"-like $^3n\pi^*$ state prefers a planar geometry. The second consequence is that the planar $^3n\pi^*$ states should have rather long CO bonds and C—C distances appropriate to a free radical $R_1R_2\dot{C}H$. Approximate equilibrium bond lengths have been determined at the "ab-initio" level [153]. In their planar conformations the two triplets are stabilized by the bond relaxation (20 kcal/mole for the $^3n\pi^*$ state and 11 kcal/mole for the $^3\pi\pi^*$ state). The equilibrium geometries for the two triplets — as well as for the two parent singlets — are similar: $R_1(C_1-C_2) \sim 1.43$ Å, $R_2(C_2-C_3) \sim 1.37$ Å and $R_3(C_3-O) \sim 1.31$ Å. Hollas [158], analyzing the intensity distribution of the main progression of the singlet $n\pi^*$ absorption band of acrolein (3860 Å), has found the following values: $R_1 \sim 1.46$ Å, $R_2 \sim 1.35$ Å and $R_3 \sim 1.32$ Å (see also Ref. [159,160]).

Table 5. Selected sets of bond lengths $R_1R_2R_3$ used to investigate the effect of the relaxation of the bonds on the $n\pi^*$ and $\pi\pi^*$ triplets of acrolein

Geometry (Å)	(0)	(1)	(2)	(3)	(4)	(5)	(6)
$R_1(C_1C_2)$	1.36	1.37	1.42	1.54	1.48	1.51	1.54
$R_2(C_2C_3)$	1.46	1.42	1.39	1.35	1.32	1.30	1.28
$R_3(C_3O)$	1.22	1.25	1.29	1.32	1.34	1.36	1.38

For each of the six systems $\{R_1, R_2, R_3\}$ listed in Table 5, the angle of twist θ_1 has been varied from 0 to 90°. The potential energy surfaces of the two triplet states are presented in Fig. 6. In order to obtain a

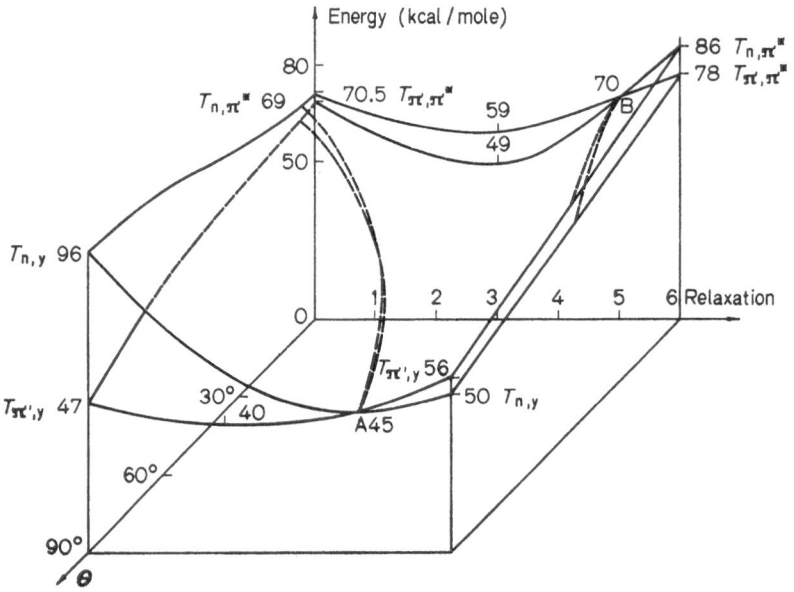

Fig. 6. Potential energy sheets of the two lowest triplet states of s-*trans* acrolein [153]

qualitatively exact description of the actual acroleinic system, the calculated $^3\pi\pi^*$ surface has been lowered by 20 kcal/mole. The two vertical $^3n\pi^*$ and $^3\pi\pi^*$ states are then nearly degenerate, as they have been found to be experimentally. The true calculated surfaces are represented in Fig. 7 (Ref. [153]). The important fact here is that these two potential energy sheets present two regions of avoided crossing. It is no longer meaningful to speak of the $n\pi^*$ or $\pi\pi^*$ triplet state but only of the lowest or highest state, the "orbital" nature of each of these states depending on the point of the sheet at which it has to be defined. As shown in Fig. 7, the vertically excited $n\pi^*$ triplet state relaxing its bonds without twisting remains a $n\pi^*$ state but, as it is allowed to twist without changing its bond lengths (geometry 0), it acquires more and more $\pi\pi^*$ character and finally becomes a pure $\pi\pi^*$ state. If the twist and the relaxation are both permitted, the $n\pi^*$ state slowly becomes the ny triplet state.

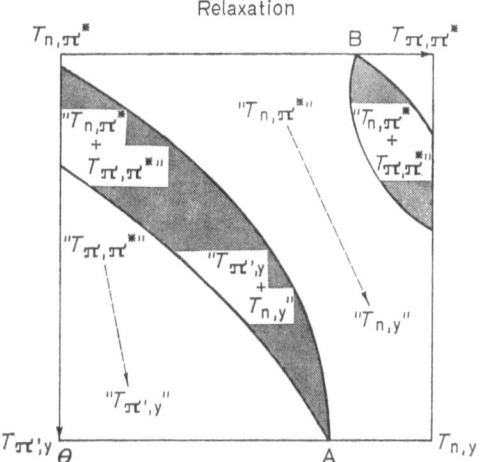

Fig. 7. "Orbital nature" of the lowest triplet potential energy sheet of s-*trans* acrolein [153]

To conclude this study of acrolein, the twist around the central C_2C_3 bond has to be examined. The experimental data on positions of zero-point levels in torsional vibration potential energy curves in acrolein are summarized in Fig. 8 [161], and the corresponding calculated curves are shown in Fig. 9 [153].

The two diagrams agree fairly well. The first important point is the existence of two rotamers of acrolein in its ground state. The first and most stable is the *trans* rotamer, the second is the *cis* form. No stable gauche form has been detected either in the present minimal basis calculation or in extended basis calculations [162], though the latter confirm the existence of a second gauche rotamer of butadiene, as predicted by Dewar and Harget [163]. The second point is that the situation in the $n\pi^*$ and $\pi\pi^*$ triplet states is the opposite, experimentally or theoretically, the *cis* form being now more stable than the *trans* form. Finally, the torsional barrier is greater in the $\pi\pi^*$ triplet state (14 kcal/mole) than in the $n\pi^*$ triplet (4.4 kcal/mole). This is readily explained in terms of MO's. The π_1 and π_2 MO's of acrolein are bonding between C_2 and C_3 and are therefore destabilized by the rotation around C_2C_3. On the other hand, π_2 is antibonding in that region and is stabilized. (The n molecular orbital remains unaffected.) In the $n\pi^*$ triplet three electrons are destabilized (those in π_1 and π_3) and two are stabilized (in π_2). The resulting net destabilization is therefore smaller than the destabilization of the $\pi\pi^*$ triplet, where three electrons are destabilized (in π_1 and π_3) but only one is stabilized (in π_2).

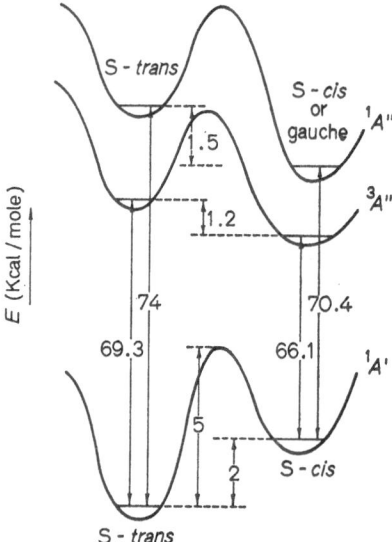

Fig. 8. Experimental data on positions of zero-point levels in torsional vibration potential energy curves of acrotein (these data include the effect of the relaxation of the bond lengths) [153]

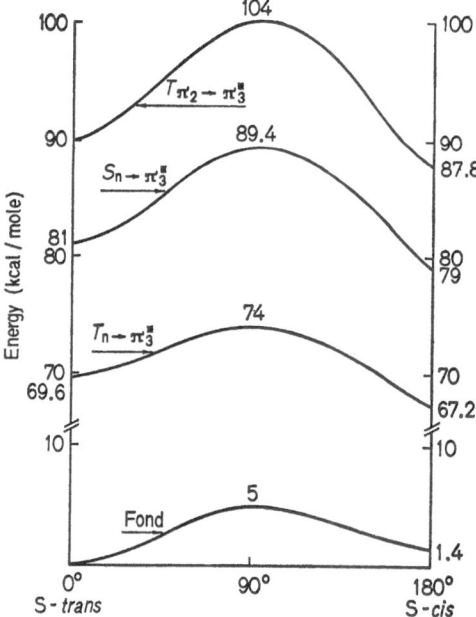

Fig. 9. Theoretical potential energy curves for the ground and excited states of acrolein in the rotation around the central bond. (These curves do not include the effect of the relaxation of the bond lengths) [153]

II.3. Dissociation of Formaldehyde and Ketones

In the foregoing examples potential energy surfaces were used simply to determine the equilibrium geometries of the low-lying excited states of conjugated molecules. Photochemical *dissociation* processes may be studied with the same tools. Formaldehyde, a prototype for carbonyl compounds, is again a simple example to start with. Its photolysis has received a renewed interest recently, as have many photolytic reactions associated with atmospheric contamination. Furthermore, the discovery of H_2CO in interstellar space [164] means that a detailed knowledge of its photochemistry is of primordial importance. Four primary processes have been postulated, one of which (IV), will be important only with veryshortwavelength radiations. The relative importance of the three remaining processes in the region of usual photolytic studies is in a

$$H_2CO \ (^1A_1) \ \xrightarrow{\ h\nu\ } \ H_2CO \ (^{1,3}A_2)$$

$$H_2CO \ (^{1,3}A_2) \ \longrightarrow \ H + HCO \qquad\qquad I$$

$$H_2CO \ (^{1,3}A_2) \ \longrightarrow \ H_2 + CO \qquad\qquad II$$

$$H_2CO \ (^{1,3}A_2) + H_2CO \ (^1A_1) \ \longrightarrow \ H_2COH + HCO \qquad III$$

$$H_2CO \ (^1A_1) \ \xrightarrow{\ h\nu\ } \ 2\,H + CO \qquad\qquad IV$$

state of confusion. The first experimental difficulty is the absence of a definite measure of the H—HCO bond strength, the results ranging from 68 to 92 kcal/mole [32]. Walsh and Berson [168] obtained a value of 87 kcal/mole and suggested that process III could be responsible for the photolysis at 365 nm, since no predissociation could occur at this wavelength if the bond dissociates at 87 kcal/mole. McQuigg and Calvert [169] set an upper limit of 85 kcal/mole (or, maybe, 81 kcal/mole) in which case it is not necessary to include III. The second experimental question is that inhibition methods [170,165] and isotope methods [169,171] give the relative quantum yield for I and II with conflicting results. At long wavelengths the latter method favors I, the former II.

The "ab-initio" study of process I, the dissociation of formaldehyde into *radical products*, was simultaneously done by Fink [90] (using Nesbet's method of symmetry and equivalence restrictions and a limited CI) and by Hayes and Morokuma [62] (using a GSMO.CI method and the "point system" in the selection of the configurations). Both sets of potential energy curves exhibit the same behavior. One is shown below (Fig. 10).

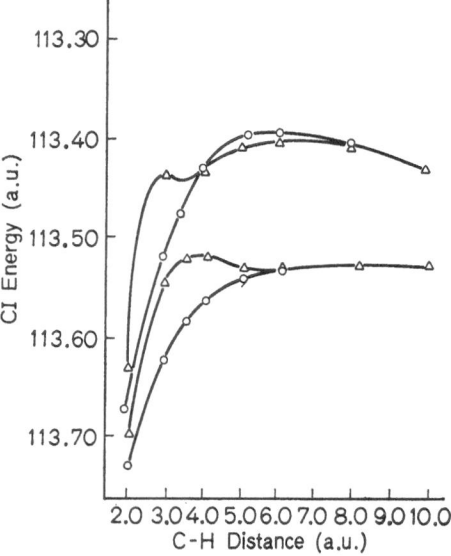

Fig. 10. Potential energy curves for the dissociation of formaldehyde into radical products. The first two singlets — the $^1A_1$ ground state and the $^1A_2$ nπ* state — are shown (○). They correlate respectively with the systems {HCO · $(^2A')$ + H · } and {HCO · $(^2A'')$ + H · }. The first two triplets — the low-lying $^3A_2$ nπ* state and the high-lying $^3A_1$ ππ* state — are also represented (△) [90]

In the dissociation process the π*$(2b_1)$MO of the nπ* state changes into a pure $1s$ orbital of hydrogen while acquiring some fairly large $3s$ character in the intermediate region (C—H distance ~ 5 au). The n$(2b_2)$MO becomes similar to the corresponding orbital of the formyl radical. The delocalization of this MO on the neighbouring carbon atom is apparent in H_2CO as well as in the HCO radical [172]. Two points are worth mentioning in the description of the states. As an illustration of our general discussion of the HF treatment at infinite separation, the SCF (without CI) $^1A_2$(nπ*) state misbehaves badly at large distances, dissociating to the same asymptotic energy as the triplet state. Since the $^1A_1$ (S_0) ground state and the $^3A_2(T_1)$ state must dissociate to the same limit, the $^1A_2(S_1)$ state should dissociate to an excited state of the formyl radical. This is ensured if $^1A_2$ is written as the mixing of the two configurations $(2b_1)^1(2b_2)^1(6a_1)^0$ and $(2b_1)^1(2b_2)^0(6a_1)^1$. The second point is that the ^{3}nπ* state reaches a saddle point for a C—H distance of ~ 3.6 au and then gradually decreases to the asymptotic value of the ground state HCO + H (The barrier is roughly 0.4 eV above this ground state). This is attributed to the crossing in this region of the π* MO (independent of the C—H distance) and a σ^* MO (which comes down rapidly when the C—H distance increases). The presence of an activation energy to dissociation in triplet acetone has been experimentally reported [173].

Assuming that the initial excitation is to the $^1A_2(n\pi^*)$ state of formaldehyde, several mechanisms may be proposed.

1. The dissociation occurs entirely on S_1. Because of the high energy of the excited formyl radical to which S_1 correlates, such a process is unlikely with usual exciting radiation ($\lambda \sim 350$ nm). It will become possible at some wavelength between 313 and 280 nm (this range corresponds to the maximum range of reported values for the H—HCO bond energy). Emission from the excited formyl radical should be observable while the H atom would have very low translational energy. If the predissociation remains predominantly to the ground state, no emission will be observed and quite energetic H atoms will be generated. However, it has been shown that energy greater than 3.66 eV ($\lambda \leqslant 338.5$ nm) is sufficient to initiate the reaction. Therefore a mechanism exists which does not require the molecule to pass to the excited states of products.

2. Intersystem crossing $S_1 \to T_1$ occurs, followed by dissociation in T_1. Lee and coworkers [174,177] have shown that the most important primary process in the photochemistry of cyclic ketones is the radiationless transition $S_1 \to T_1$. If this is the case for formaldehyde, the experimental observation that a nonplanar dissociative pathway is favored, in which the leaving H atom has a velocity component perpendicular to the plane of the HCO fragment [178], may be explained in terms of the "static" potential energy curves alone. However, it has also been found that the lifetime for the decay process leading to the radical products from S_1 is $< 10^{-11}$ sec [178]. Intersystem crossing is not expected to be that fast. In cyclic ketones its rate is of the order of 5×10^{-8} sec [179]. It has been shown that the rate of radiationless processes increases appreciably with increasing vibrational energy [180,181] (whereas radiative decay from S_1 remains unchanged), but the rate enhancement is not sufficient to reconcile the two values.

3. Internal conversion occurs from S_1 to a highly excited vibrational level of S_0. On the assumption that processes that conserve spin occur faster than those that do not ($S_1 \to T_1$), this path might explain the great speed of I. However, the reverse seems to be true in the case of ketones [182,187]. The calculations show that the dissociation pathway of the ground state is planar [62]. The experimental nonplanarity can no longer be explained in terms of "static" potential energy

41

curves but solely in terms of dynamical effects. Due to the nonplanarity of S_1, some of the transferred energy will be localized in the out-of-plane mode of vibration of S_0. This, and the mass ratio of the two (H and HCO) components, might account for the nonplanar dissociation [62].

The dissociation of H_2CO into "*molecular products* (process II) has also been investigated by Fink [90]. A representative path is calculated by modifying the ground-state geometry in the following ways: the two hydrogens are bent out of the plane (by 31°), the HCH angle is reduced to such an extent that the distance between the hydrogen atoms is the value found in the hydrogen molecule (these modifications will over-emphasize the steep rise in energy close to the starting geometry of form-aldehyde. The dissociation curves are obtained by symmetrically removing the two hydrogens from the carbon atom (Fig. 11). The highest doubly

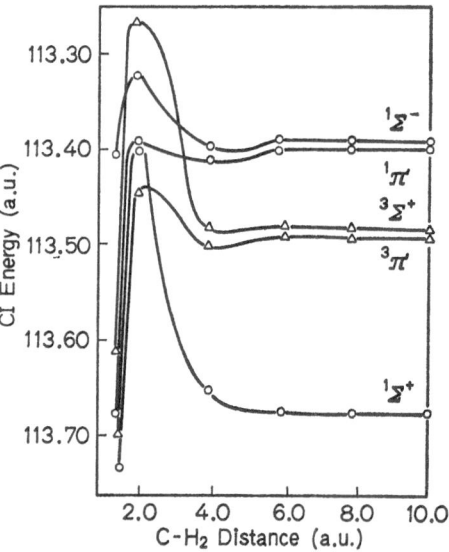

Fig. 11. Potential energy curves for the dissociation of formaldehyde into molecular products (H_2 + HCO). The first three singlets — the $^1A_1$ ground state, $^1A_2$ $n\pi^*$ state and $^1A_1$ $\pi\pi^*$ state — are shown ($\bigcirc$). The first two triplets — $^3A_2$ $n\pi^*$ state and $^3A_1$ $\pi\pi^*$ state — are also represented ($\triangle$) [90]

occupied MO ($1b_1$) then becomes the bonding orbital of the hydrogen molecule. The singly occupied orbitals $2b_1(n)$ and $2b_2(\pi^*)$ correlate with π-type levels of the CO fragment. In contrast to the radical dissociation, the proper behaviors are immediately obtained at the simple SCF (without CI) level because the H_2 fragment is a closed-shell subsystem. The $^{1,3}A_2$ states then correlate with the $^{1,3}\Pi$ excited states of carbon monoxide. Something less than 10 kcal/mole being required, the dissociation may be thermodynamically possible by irradiation with near-ir wavelengths. The first excited state of $CO(^3\Pi)$ being ~ 6 eV above the ground state, the first triplet dissociation does not play any role until a wavelength of over 185 nm is used. Therefore, for most of the available photolytic energies, the extra energy will be trapped inside the two ground-state fragments.

To conclude this study of formaldehyde, let us briefly consider the possibility of competition between processes I and II. First, one hydrogen is easily abstracted from the ground-state formaldehyde since there is no activation barrier in the ground-state surface. Second, the possibility of a Herzberg type-II predissociation process giving molecular products (while forbidding the dissociation into radicals until higher energies) is clearly evident, as is the possibility of a second excited state becoming rapidly operative for the radical dissociation but not for the molecular fragmentation. If a molecule is excited above the H—HCO bond-dissociation energy, it can transfer to the ground-state surface. It will then find itself inside a bound well as far as process II is concerned and will have to undergo barrier tunneling to dissociate. Conversely, this transfer leads easily to the direct dissociation I. There will be a smaller, temperature-independent pre-exponential factor for II than for I. If, on the other hand, the excitation is smaller than the H—HCO dissociation energy, I will not be possible but II will still be possible, though at a lower rate, via tunneling.

The α cleavage of saturated ketones [188,193] has also been examined theoretically [194]. This example and several others — some of which will be described in the following sections of this article — serve to illustrate a simple yet realistic interpretation of photochemical reactions in terms of surface crossings [195] or surface touchings between ground state and excited singlet and triplet states. Salem's model throws additional light on the intimate mechanisms of photoreactivity by (a) clearly defining the number and electronic characteristics of the radical centers created during the reaction and (b) relating the symmetry of the diradical product and the multiplicity of the photoreactive state. Still more important is the fact that this pioneering study is the basis for a detailed classification of numerous photochemical reactions. The proper treatment of the diradical obtained in the first stages of the dissociation employs either a ground-state closed-shell calculation followed by a 3×3 CI treatment (including the ground state, the lowest singlet excited state and the corresponding doubly excited state), or an open-shell calculation based on Nesbet's hamiltonian (46). For a given state, the energy chosen is the lowest of the two possible roots. Experience shows that the open-shell method always gives the lowest energy for the diradical-like $^{1,3}n\pi^*$ states whereas the GSMO-CI method is better suited for bonded

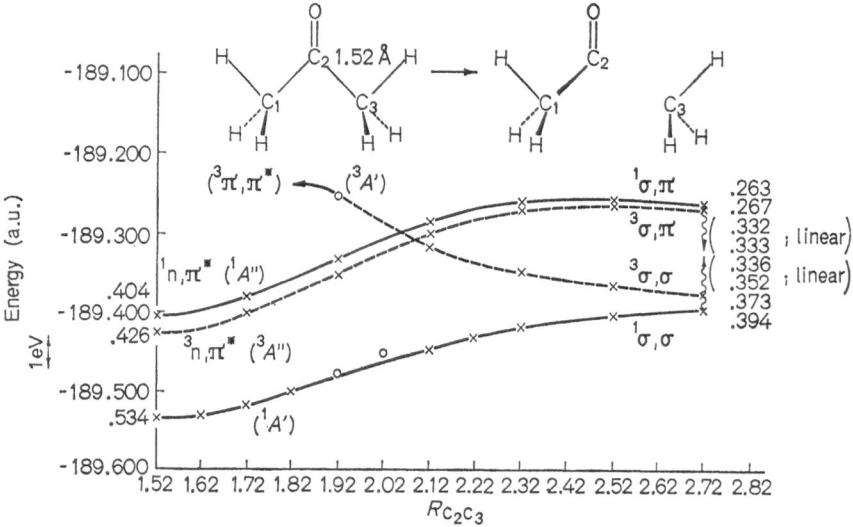

Fig. 12. Energy surfaces for α-cleavage of acetone [194]

or zwitterionic states. Fig. 12 represents the potential energy curves for various excited states of acetone when this molecule dissociates to give a *bent* acetyl radical. The behavior of the $n\pi^*$ and $\pi\pi^*$ triplet curves allows us to explain the barrier which appeared in the curve of the $^3A_2$ state of formaldehyde. In a planar process the two triplet states belong to different irreducible representations of the C_s point group: $A'(^3\pi\pi^*)$ and $A''(^3n\pi^*)$ and cross. In a non-planar dissociation path like the one investigated by Fink the symmetry restriction is lifted. The barrier in the lowest state is the manifestation of the resulting avoided crossing. In the dissociation of acetone to a bent acetyl radical the crossing region ($C_2C_3 = 2.10$ Å) is the best place for radiationless transitions to occur: intersystem crossing from the singlet $n\pi^*$ to one or both of the available triplet manifolds, or internal crossing within these triplet manifolds. This extremely favorable intersystem crossing region must be relevant to the participation of both the singlet and triplet $n\pi^*$ states in the cleavage of alkyl ketones in solution [184]. The triplet state is even found to be more than a hundred times more reactive than the singlet[182,183] with the consequence that the investigation of the α cleavage of a variety of cyclopentanones and cyclohexanones in solution indicates that the products come solely from the triplet $n\pi^*$ state [186,187]. In the case of a dissociation path leading to a *linear* acetyl radical (the corresponding asymptotic values are given on the right side of Fig. 12), there is a near touching of the two singlet (ground and $^1n\pi^*$) and of the two triplet ($^3n\pi^*$ and $^3\pi\pi^*$) states. If the excited molecule travels on the $^1n\pi^*$ state throughout, it then relaxes to its linear equilibrium geometry from which internal conversion to a linear ground state is quite easy.

The assumed conservation of a plane of symmetry in the cleavage of acetone may allow us to rationalize the theoretical results in terms of a state correlation diagram. Relative to this plane of symmetry one electron has either σ or π symmetry, and the correlation between states involves a simple count of the electrons of each type. In the α cleavage of acetone the acetyl radical has two available low-lying states. The first

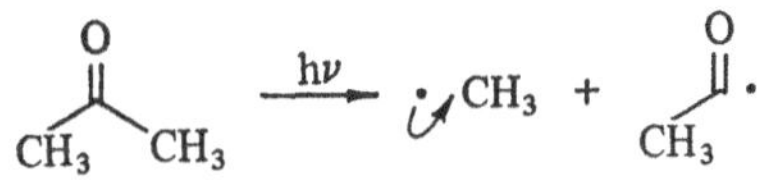

is a σ radical with the odd electron in an in-plane σ orbital. This radical has two π electrons (in the π bonding CO.MO) and 3σ electrons (the two lone-pair electrons on the oxygen atom and the odd electron resulting from the homolytic cleavage, located on the carbon atom). The second state is a linear excited state with the odd electron in the π system. There are therefore 3π electrons in this structure (the three dots in (b)) and two

(a) σ radical   (b) π radical

remaining σ electrons, which may be represented as shared by the two C and O atoms in an in-plane delocalized σ bond. For the formyl radical the bent σ radical $^2A'$ lies 1.1 eV below the linear π radical ($^2A''$) [196,197]. The σ acetyl radical is also found by "ab-initio" calculations 1.25 eV below the acetyl π radical. If we restrict the count of the electrons to the two carbonyl π electrons, the two oxygen lone-pair electrons, and the two electrons of the CC bond which is cleaved, the correlation diagram that schematically represents the calculated potential energy curves of Fig. 12 is easily obtained (Fig. 13). The important features of this diagram are the

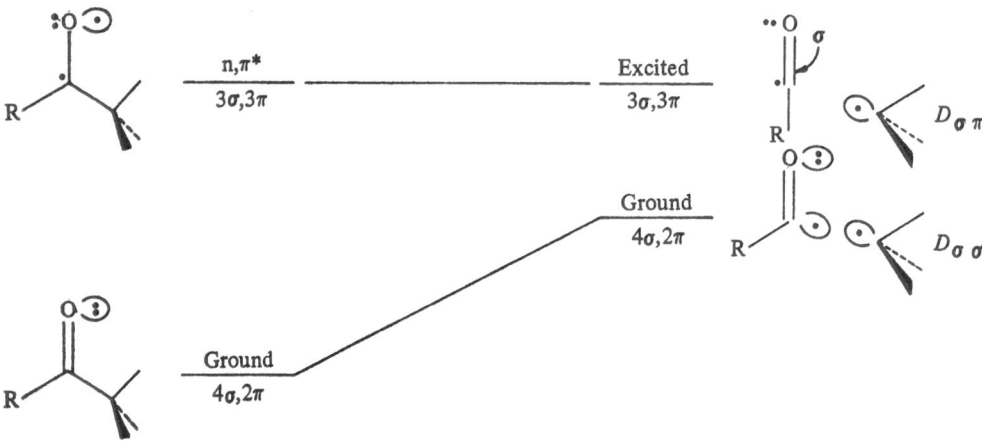

Fig. 13. State correlation diagram for α-cleavage of alkanones [194]

following: (1) there is no crossing of the two singlet surfaces, rather there is a near touching of these surfaces since the two linear π and bent σ acetyl radicals — and therefore the two corresponding diradicals $D_{\sigma\pi}$ and $D_{\sigma\sigma}$ — are very close in energy; (2) the triplet associated with the lowest $D_{\sigma\sigma}$ radical — and having nearly equal energy — is symmetric and must correlate with the lowest symmetric triplet of acetone, i.e. the $\pi\pi^*$ triplet. As we have seen in the various calculations on formaldehyde, this triplet is higher than the [1,3] $n\pi^*$ states and therefore comes down, crosses the singlet and triplet $n\pi^*$ curves and, acquiring more and more $\sigma\sigma$ character, finally correlates with the triplet $D_{\sigma\sigma}$ diradical.

This scheme, which is completely confirmed by the calculations on formaldehyde and acetone, may be extended to conjugated enones. The introduction of one (enones) or two (dienones) C=C double bonds will result in the greater stabilization of the linear π radical relative to the bent σ radical. With acrolein, the relative position of the two "acrolyl-2" radicals is still ambiguous. A minimal basis-set SCF calculation favors the σ-radical ground state by 0.8 eV whereas an INDO calculation yields a π-radical ground state, 0.87 eV below the bent radical [198]. In the dienone case the π radical is almost certainly the more stable of the two structures, and the correlation diagram for cyclohexadienone is as represented in Fig. 14 and is characterized by a crossing of the two singlet curves at the very end of the dissociation pathway. The two surfaces are nevertheless again in a "touching" situation. As for the triplets, the $n\pi^*$ state will closely follow the parent singlet, though slightly below it. The $^3\pi\pi^*$ state on the other hand is, like the π radical,

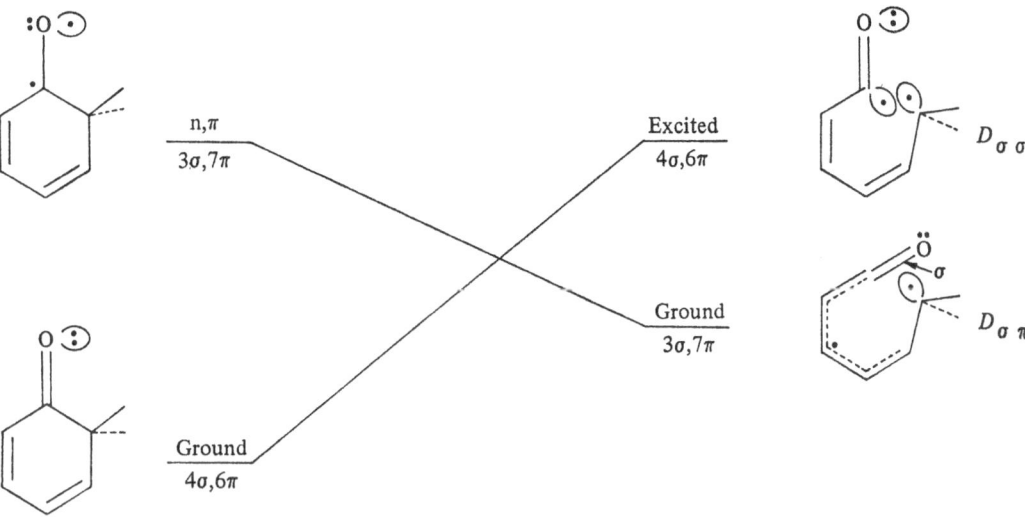

Fig. 14. State correlation diagram for α-cleavage of cyclohexadienones [194]

strongly stabilized by conjugation with the C=C double bonds and will certainly lie below the $n\pi^*$ states. Since this state correlates with the symmetric excited triplet $D_{\sigma\sigma}$, its rising surface will cross those of the two $n\pi^*$ states. The crossing region will be a region of intense radiationless transitions. The two diradicals, being structurally very different, will react differently. The $D_{\sigma\pi}$ diradical might lead to a dienyl-ketene by simple rotation of the methylene radical center [199] whereas the $D_{\sigma\sigma}$ diradical can form a bicyclic product, the bicyclo[3.1.0] hexenone.

$D_{\sigma\pi}$ Dienyl—ketene

$D_{\sigma\sigma}$ Bicyclo [3.1.0] hexenone

47

II.4. Ring-opening of Azirines

Substituted azirines add photochemically to olefins and ketones to yield five-membered rings of the type of pyrroline (($X=CR_1R_2$) or 3.oxazoline ($X=O$) [200,203]. The primary $n\pi^*$ excitation, like the $n\pi^*$ excitation of ketones, leads to the cleavage of a C—C single bond α to the site of

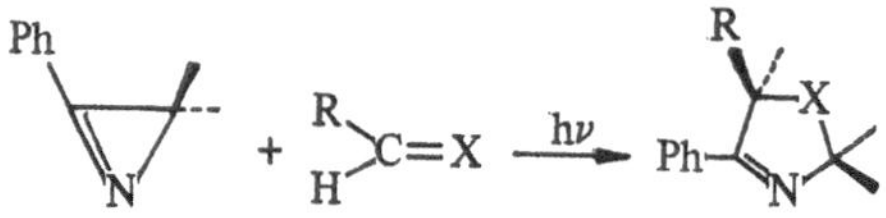

excitation. Three radical centers are then created (a π center with three electrons and two centers resulting from the homolytic cleavage of the CC bond). If the count of electrons includes the two C=N π electrons, the two nitrogen lone-pair electrons, and the two electrons of the CC bond, we may expect the primary photoproduct to be one of the two low-lying neutral radicals $D_{\sigma\sigma}$ (A' symmetry due to 2π and 4σ electrons) and $D_{\sigma\pi}$ (A'' symmetry due to 3π and 3σ electrons). In $D_{\sigma\pi}$ the three dots on the C and N atoms are originally the three π electrons and σ denotes the in-plane "ketiminoïd" part of the double bond. This simple situation is however complicated by the possibility of ionic resonance

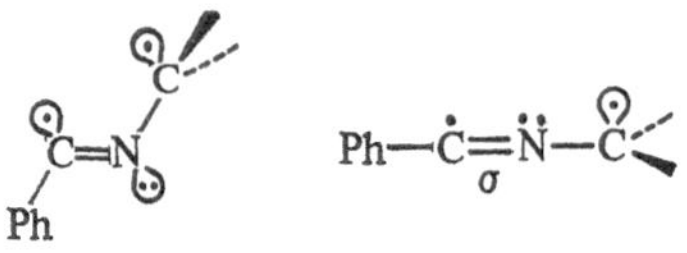

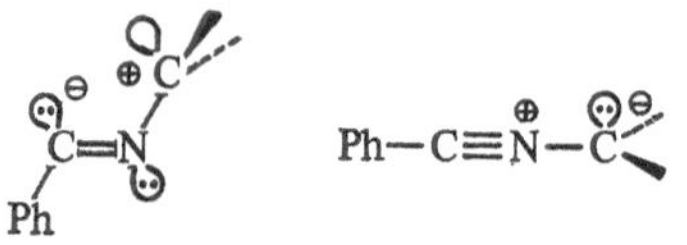

structures of the 1.3 dipolar type. Two of them, I and II, were postulated experimentally [200,203]. However, it is known that, when allowed by symmetry, a molecule with a broken bond can have both covalent (neutral) and ionic character, the ground and the excited states of this molecule being mixtures of diradical and zwitterionic components [204, 205]. In the present case the $D_{\sigma\sigma}$ diradical structure and the ionic structure I are two resonance forms (with 4 σ and 2 π electrons) occurring simultaneously in the description of a bent singlet state. The linear ionic structure II is also a 4 σ ,2 π form and, as such, occurs in the description of the same, but now linear singlet state, together with the linear form of the diradical $D_{\sigma\sigma}$ (see Fig. 15). By analogy with the formyl

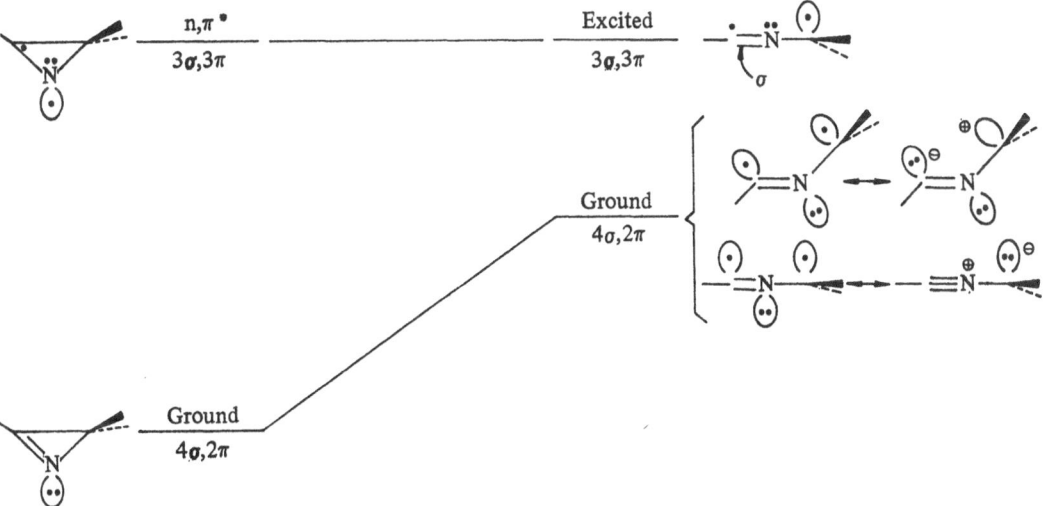

Fig. 15. State correlation diagram for the ring-opening reaction of azirines [194]

radical, the σ imine radical moiety should be more stable then the π imine moiety. The $D_{\sigma\sigma}$ diradical is therefore expected to be the most stable and the correlation diagram follows easily. This qualitative diagram has been confirmed by "ab-initio" calculations (Fig. 16). The calculated potential energy curves display two interesting features. If the ring is opened to an intermediate bent geometry (<CNC=110°) the $^1A'$ ground state is destabilized while, at the same time, the *second* triplet state (probably $^3\pi\pi^*$ but possibly $^3n\sigma^*$, depending on which one of these two states is the lowest) comes down rapidly and becomes the ground state $^3\sigma\sigma$ of the bent molecule, 1.5 eV below the $^1\sigma\sigma(A')$ state. The two states are essentially diradical-like in this region. When the CNC angle opens up further toward the linear geometry, the triplet state is destabilized and $^1\sigma\sigma$ is stabilized. This singlet state acquires closed-shell zwitterionic character and the calculated net charges — C_3:

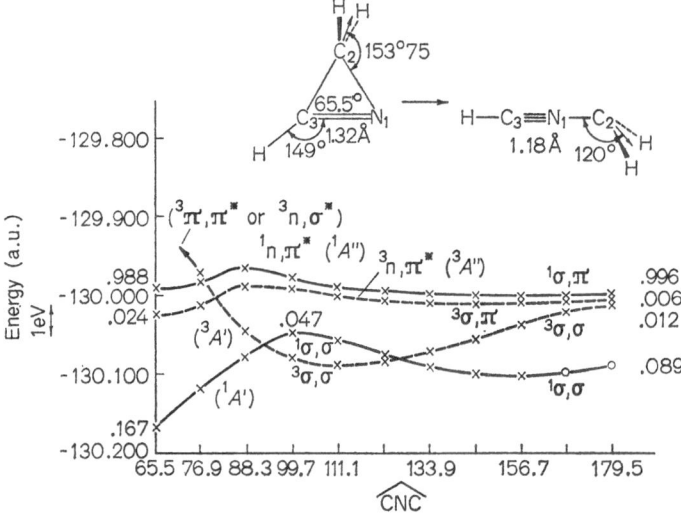

Fig. 16. Energy surfaces for the ring-opening of 2 H-azirine [194]

0.37, N_1: 0.11 and C_2: -0.46 — are in good agreement with a 1.3 dipolar structure similar to II, the positive charge being shared by the carbon (C_3) and nitrogen atoms. Depending on its geometry when it collides with a second molecule, photochemically excited aziridine may then react as a bent $^3\sigma\sigma$ diradical or a linear $^1\sigma\sigma$, 1,3 dipolar, zwitterionic species.

II.5. Hydrogen Abstraction by Ketones [194]

"Ab-initio" calculations will now be used to investigate photochemical bimolecular reactions. The two reacting molecules are considered as a single supermolecule. For example, the electronic states of the pair of radicals resulting from the intermolecular abstraction of a hydrogen by a ketone [188,206,211] may be considered as the electronic states of the super-diradical.

$$\underset{\underset{R}{\overset{\displaystyle O}{\overset{\|}{C}}}{\diagdown}R'}{} + H-R'' \xrightarrow{h\nu} \cdot R'' + \underset{\underset{R}{\overset{\displaystyle O-H}{\overset{|}{C\cdot}}}{\diagdown}R'}{}$$

The excited state responsible for the reaction can be either the singlet or the triplet $n\pi^*$ excited state of the ketone [212,216].

Let us consider the correlation diagram linking the ground and lowest excited state of both reactants and products in the model case where the abstraction of a hydrogen atom of methane is accomplished by formaldehyde. The determination of the global symmetry of the states will be extremely simple if we assume that the abstraction occurs in the plane of the formaldehyde molecule, this plane remaining a plane of symmetry of the entire system throughout the reaction. The relevant electrons to be counted are the two π electrons of the CO bond, the two lone-pair electrons originally on the oxygen atom, and the two electrons of the σ CH bond to be destroyed. As shown on the left of the correlation diagram (Fig. 17), the ground and first $n\pi^*$ excited states of the reactants are characterized by (4σ, 2π) and (3σ, 3π) electrons, respectively. (The π

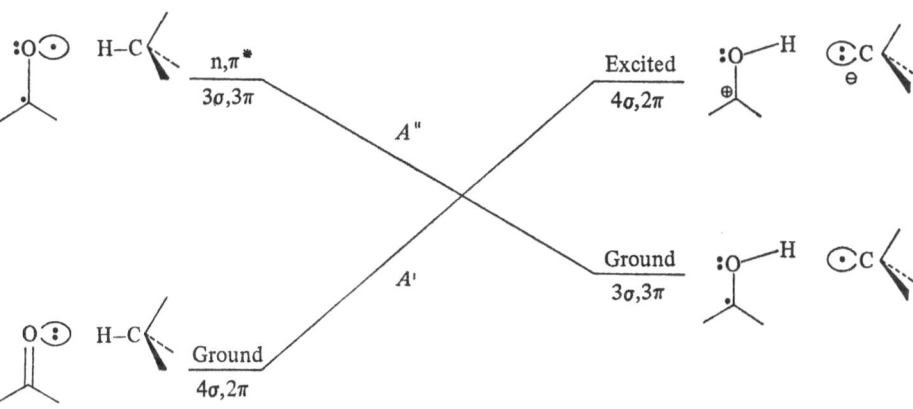

Fig. 17. State correlation diagram for hydrogen abstraction by ketones [194]

electrons are represented as part of the double bond in the formaldehyde ground state and by the three dots in the $n\pi^*$ state). The ground state of the primary products is a diradical state with 3σ and 3π electrons. Such a state is always accompanied by two zwitterionic states [204,205,217,218], the lowest one (4σ and 2π electrons) being displayed here. The correlation diagram shows that the ground-state surface of the reactants (A' symmetry) correlates with the ionic excited state of the products and crosses the excited-state surface of the reactants (A'' symmetry) which goes down to the diradical ground state of the products. This crossing is strictly allowed since the two curves have opposite symmetries but, in practice, will be weakly avoided because the H abstraction need not occur in a coplanar fashion [219]. The important feature of this diagram is that the reaction leads from the excited state of the reactants to the ground state of the products in a perfectly *adiabatic* manner.

Fig. 18 represents the results of the "ab-initio" study of this reaction. The carbonyl bond length and the COH$_6$ angle were allowed to vary

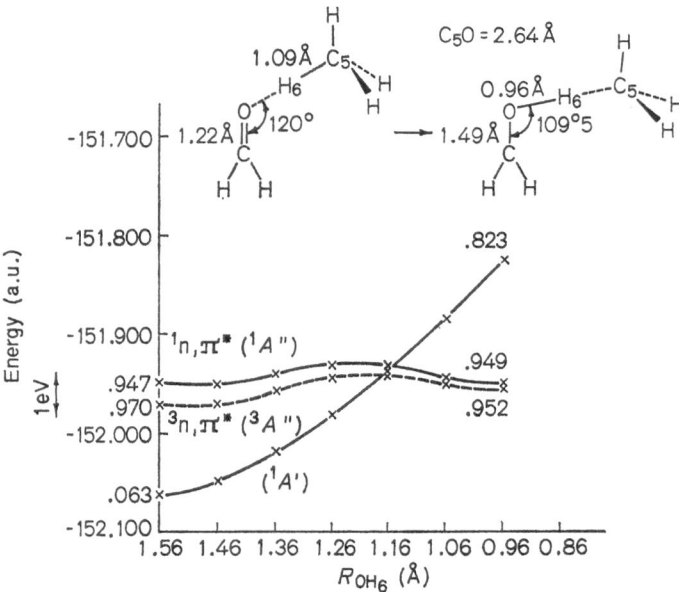

Fig. 18. Energy surfaces for the hydrogen abstraction in the methane-formaldehyde system [194]

linearly from their initial values (1.22 Å, 120°) to the final ones (1.43 Å, 109° 5), the C$_5$O distance being kept constant. First of all, the predicted crossing of the two singlet potential energy curves is confirmed .This crossing takes place for a OH$_6$ distance of 1.16 Å, *i.e.* near the end of the adiabatic process. Second, energy barriers are found in the singlet (8 kcal/mole) and triplet (17.5 kcal/mole) nπ* states. The experimental activation energy for abstraction in the triplet state is found in the range 4.2—7.1 kcal/mole [220]. The theoretical value is 10 kcal/mole larger, but no geometry optimization has been performed on the excited states. Finally, the energy difference between the singlet ground states of reactants and products is 72 kcal/mole, in excellent agreement with thermochemical estimates (74 kcal/mole) where losses of 85 kcal/mole (disappearance of the CO π bond) and 99 kcal/mole (breaking of the CH bond) are partly compensated by the gain resulting from the newly formed OH bond (110 kcal/mole).

II.6. Addition of Excited Ketones to Olefins

Excited ketones add to electron-poor olefins via the CO π orbital and to electron-rich olefins via the n oxygen lone pair [208,209,221,224]. We shall consider here a model system for the second family of reactions and assume that ethylene approaches formaldehyde in such a manner that the ethylene π system lies in the plane of the formaldehyde molecule. This plane is therefore the plane of symmetry of the supermolecule.

The electrons to be accounted for are the two π electrons of the C=O bond, those of the C=C π bond, and the two lone-pair electrons of formaldehyde. The correlation diagram is easily obtained, the diradical ground state of the products being once again accompanied by the lowest of the two possible zwitterionic states. As shown in Fig. 19, the prominent feature of this diagram is the allowed crossing between the two singlet

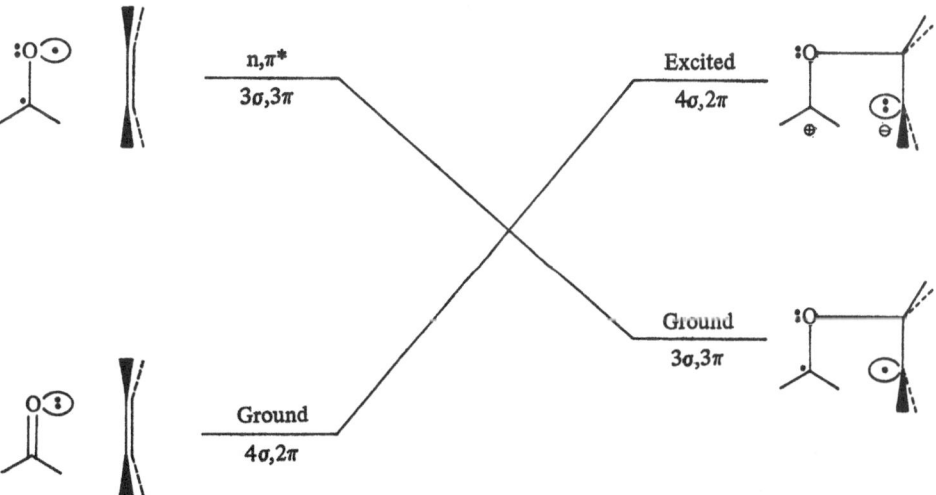

Fig. 19. State correlation diagram for the (n orbital) addition of ketone to olefins [194]

potential energy surfaces, each ground state (reactants, products) correlating with the lowest excited state (products, reactants). The corresponding "ab-initio" potential energy curves are shown in Fig. 20

53

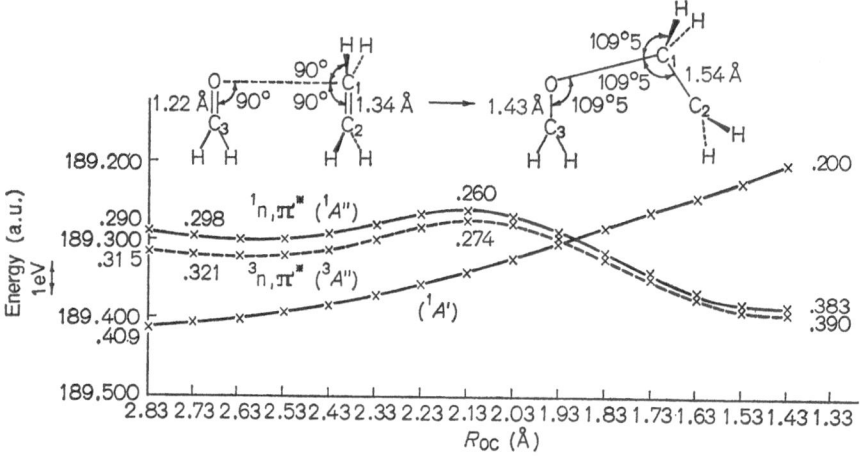

Fig. 20. Energy surfaces for (n orbital) addition of formaldehyde to ethylene [194]

and, aside from the clear crossing of the singlet curves, two characteristics are worth mentioning. First the singlet–triplet (nπ*) separation decreases slowly from left (isolated ketone: 0.7 eV) to right (diradical products: 0.2 eV) as the diradical character increases in the two states. The same phenomenon might have been noted in the hydrogen abstraction by ketones (Fig. 18). Second, at a distance of 2.6 Å both nπ* states display a shallow minimum whose depth is ~5 kcal/mole relative to the isolated molecules. This arises from the initial stabilization of excited formaldehyde when the CO bond is lengthened, a relaxation process mentioned in the first part of this article. This artefact would have been removed if we had started with fully relaxed nπ* states, so this minimum does not appear to be related to the observed excited-state complex, which seems to be formed prior to appearance of the diradical [225].

II.7. Methylene Photochemistry

In its linear conformation methylene (carbene) has the configuration

$$1\,\sigma_g^2\,2\,\sigma_g^2\,1\,\sigma_u^2\,1\,\pi_u^2$$

which, because of the twofold degeneracy of the $1\,\pi_u$ orbital, leads to three low-lying electronic states ($^3\Sigma_g^-$, $^1\Delta_g$ and $^1\Sigma_g^+$). When the HCH angle deviates from linearity the $1\pi_u$ degeneracy is lifted, and consequently the $^1\Delta_g$ state splits into an $^1A_1$ and a $^1B_1$ state:

$$^1\Sigma_g^+ \longrightarrow {}^1A_1^* \qquad 1\,a_1^2\;2\,a_1^2\;1\,b_2^2\;1\,b_1^2$$

$$^1\Delta_g \bigg\langle \begin{array}{l} {}^1B_1 \\ {}^1A_1 \end{array} \qquad \begin{array}{l} 1\,a_1^2\;2\,a_1^2\;1\,b_2^2\;3\,a_1\,1\,b_1 \\ 1\,a_1^2\;2\,a_1^2\;1\,b_2^2\;3\,a_1^2 \end{array}$$

$$^3\Sigma_g^- \longrightarrow {}^3B_1 \qquad 1\,a_1^2\;2\,a_1^2\;1\,b_2^2\;3\,a_1\,1\,b_1$$

The two prominent MO's in these configurations are the $3a_1$ in-plane MO (which is roughly a carbon sp_2 hybrid) and the $1b_1$ MO (which is a pure p_π orbital perpendicular to the CH_2 plane). According to Walsh's rules, the $^3B_1$ triplet state should be almost linear whereas the $^1A_1$ closed-shell should be bent. Note that, due to the near degeneracy of the $3a_1$ and $1b_1$ MO's, there should be a large mixing of the two $^1A_1$ states for large bond angles.

The $^3B_1$ state has indeed been found to be the ground state of methylene by optical spectra [226-228] and ESR [229-230]. The equilibrium angle for the singlet A_1 (102° 4 [227]) and the triplet B_1 (136°±8° [228,230]) states is, as expected, very different. The energy gap between these two states is rather uncertain. Herzberg estimates an upper limit of 23 kcal/mole, but thermochemical experiments [231,232] (photolysis of ketene) and the study of the deactivation of the singlet and triplet species [233] give much smaller values (2.7 kcal/mole [231], 1–2 kcal/mole [232] and 0.6 kcal/mole [233]). Theoretical calculations give a value in the range 10–40 kcal/mole [234]. At the SCF level, for example, this gap is 32–40 kcal/mole if the basis set does not contain polarization functions, and $\sim$25 kcal/mole if it does (d functions on the carbon atom, and p functions on the hydrogens). CI treatments on the other hand lower this value to 20–25 kcal/mole (without polarization functions) and 12 kcal/mole (with polarization functions).

This simple molecule may serve to illustrate the calculation of the correlation energy in a triplet state [234]. This energy, the difference between the exact and the Hartree-Fock energies, has been calculated using formula (49). First the delocalized MO's have been transformed into localized ones according to the localization criterion of Boys [235]. The $^1A_1$ state then becomes $|K^2 l^2 r^2 n^2|$ where K denotes the carbon K shell, l and r the right and left CH bonds, and n the "sp_2" lone pair. A complete localization cannot be achieved for the $^3B_1$ state since one cannot mix singly and doubly occupied MO's. The localized wave function for the $^3B_1$ state is, at best, written as $|K^2 l^2 r^2 3a_1 1b_1|$ where $3a_1$

is almost identical to the lone-pair orbital n. The characteristics of the intra- and interpair correlation energies may be listed as follows:

1. The intrapair terms $\varepsilon(l^2)$ and $\varepsilon(r^2)$ for the CH bonds are independent of θ, the HCH angle. $\varepsilon(n^2)$ is, however, very sensitive to changes in θ because of the modification of hybridization of this orbital with θ (a pure p orbital at $180°$ becoming a sp_2 hybrid at $120°$) and of the quasi-degeneracy of the $3a_1$ and $1b_1$ orbitals for large angles θ.
2. The interpair correlation energies $\varepsilon(i,j)$ $i \neq j$ depend very strongly on the differential overlap of the two localized orbitals i and j. For example, the absolute value of $\varepsilon(l,r)$, $|\varepsilon(l,r)|$ becomes larger with decreasing θ and, at the same time, $|\varepsilon(l,n)|$ for $^1A_1$ and $|\varepsilon(l,3a_1)|$ and $|^3\varepsilon(3a_1, 1b_1)|$ for $^3B_1$ become smaller, since n (or $3a_1$) is pushed away from the carbon atom. $\varepsilon(l, 1b_1)$ depends less significantly on θ, since the distance between the centers of gravity does not depend on θ.
3. The $1b_1$ orbital, being empty in $^1A_1$, is still available for excitation in the CI process (remember that the ε's are calculated using CI schemes). As a consequence, the absolute values of $\varepsilon(l^2)$ and $\varepsilon(l,r)$ are appreciably larger in $^1A_1$ than in $^3B_1$.

As is clear from these remarks, the absolute value of the total correlation energy $|E_C|$ is, as expected, much larger in $^1A_1$ than in $^3B_1$, the difference being 15 kcal/mole. As a consequence, the gap between the two states is further reduced to 9 kcal/mole (Fig. 21). This difference in

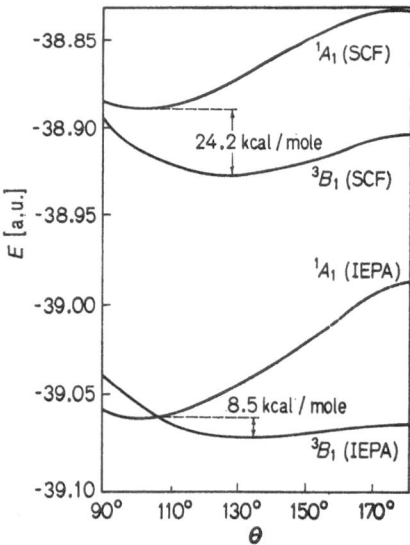

Fig. 21. SCF and IEPA potential energy curves for the two lowest states of CH_2 234)

correlation energies is due to (a) the large value of $\varepsilon(n^2)$ in $^1A_1$, which has no counterpart in $^3B_1$, and (b) the larger values of $\varepsilon(l^2)$ and $\varepsilon(l,r)$ mentioned in (3) above. Staemmler has also investigated the angular dependence of the correlation energies. $E_C(^1A_1)$ is almost θ-independent, since the changes in $\varepsilon(l,r)$ and $\varepsilon(l,n)$ roughly cancel each other (Fig. 22). $|E_C(^3B_1)|$, however, shows a strong increase with increasing θ due to the simultaneous increase of $\varepsilon(l,3a_1)$ and, to a lesser extent, of $\varepsilon(l, 1b_1)$

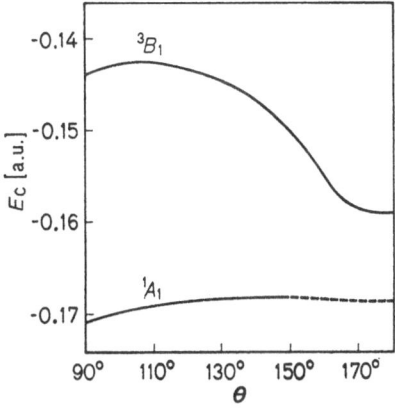

Fig. 22. Angular dependence of the correlation energy E_c for the two lowest states of CH_2 [234]

and $^3\varepsilon(3a_1, 1b_1)$. As a result, the difference between the total energy of the molecule for $\theta = 180°$ and $\theta = 135°$ is lowered appreciably (12 kcal/mole at the SCF level to 3.5 kcal/mole when the correlation is taken into account) and the equilibrium angle is enlarged (from 128° 5 to 134° 2).

Let us turn our attention toward the three major reactions of triplet carbenes in the presence of hydrocarbons: addition to olefinic C=C bonds, abstraction of hydrogen from CH bonds, and insertion into alkane CH bonds.

1. *The triplet addition of carbenes to C=C double bonds*[236,238] has been theoretically studied by Dewar and collaborators [239] using the elaborate semi-empirical MINDO/2 method which has been shown to give satisfactory estimates of molecular geometries and heats of formation for triplet states [240,241]. As shown in Fig. 24, the potential energy surface for this reaction presents three different valleys. The carbene initially approaches the ethylene symmetrically as though it were about to form a π complex. Before doing so, it veers to one side or the other,

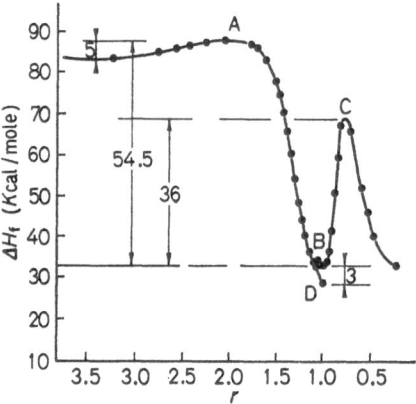

Fig. 23. Potential energy curve for the addition of triplet carbene to ethylene [239]

the two (right or left) valleys conducting directly to the products B and B'. An activation energy of 5 kcal/mole characterizes this process. The product, the trimethylene diradical (the triplet state of cyclopropane being unstable) has two conformations, B ($\Delta H_f = 32.8$ kcal/mole) which is a face-to-face diradical FF[89] (the terminal CH_2 groups are *perpendicular* to the plane of the carbons), and D ($\Delta H_f = 29.8$ kcal/mole) which has an edge-to-edge structure EE[89] obtained from FF by a 90° rotation

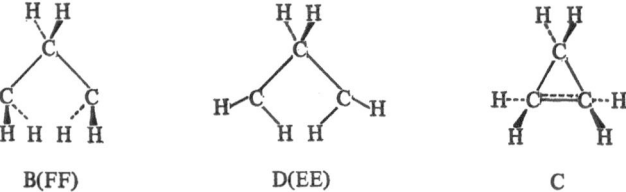

of the two CH_2 groups (which are therefore *in* the plane of the carbons). It is experimentally confirmed that triplet carbene adds to the double bond in a nonstereospecific manner, indicating the formation of an intermediate trimethylene (triplet) diradical [242,246]. The third valley (dotted line in Fig. 24) connects B and B' through the symmetric π complex C. The scrambling of the two methylene groups necessitates a (calculated) activation energy of 36 kcal/mole. (This was expected, since the π radical contains one electron in an antibonding MO, and explains the failure of such radicals to undergo Wagner-Meerwein rearrangements [247]). Fig. 23 shows the behavior of the energy of the reacting system along the three valleys of the potential energy surface.

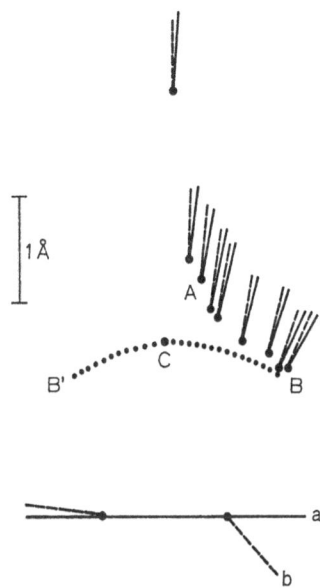

Fig. 24. Geometry of approach of triplet methylene to ethylene. The dots indicate successive positions of the methylenic carbon atom. The carbene is lying in a plane perpendicular to the paper. Lines *a* and *b* indicate the initial and final position of the hydrogen atoms at the central carbon atom of the resulting $\cdot CH_2CH_2CH_2\cdot$ diradical. The dotted lines indicate the reaction path for rearrangement of the diradical through the π complex C [239]

2. *The hydrogen abstraction* by a methylene [248,250] is highly spin-specific. In the gas phase it occurs essentially in the triplet ground state [251,254] (singlet methylene does, however, abstract halogen atoms [255]). Let us consider a planar model process in which the cleaved CH bond of methane remains in the plane of the methylene group, the displaced hydrogen atom staying colinear with the two carbon atoms [194].

$$\begin{array}{c}
\text{H} \\ \diagdown \\ \text{H}
\end{array} C + H - C \begin{array}{c} \diagup \text{H} \\ \diagdown \text{H} \\ | \\ \text{H} \end{array} \longrightarrow \begin{array}{c} \text{H} \\ \diagdown \\ \text{H} \end{array} C - H + \cdot C \begin{array}{c} \diagup \text{H} \\ \diagdown \text{H} \\ | \\ \text{H} \end{array}$$

59

The four states on the left of the correlation diagram (Fig. 25) correspond to the four configurations ($^3B_1$, $^1A_1$, $^1B_1$, $^1A_1$) mentioned at the

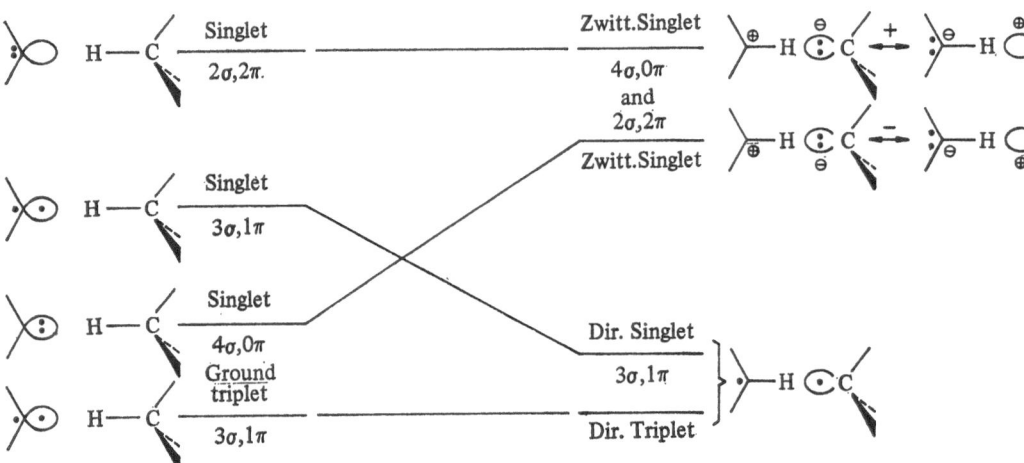

Fig. 25. State correlation diagram for hydrogen abstraction by methylene [194]

beginning of this paragraph. On the right the electronic states of the supermolecule ($\cdot CH_3 + \cdot CH_3$) correspond to the two (singlet and triplet) diradical-like structures and the two zwitterionic states represented as in-phase and out-of-phase combinations of the two possible ionic structures. These have the same symmetry, the first containing 0π and 4σ electrons, the second 2π and 2σ. This correlation diagram shows that the abstraction may take place either in the triplet ground state or in the parent singlet (of similar configuration) which, however, is the *second* excited singlet state. The reaction is clearly forbidden in the *first* singlet state. Ab-initio calculations show that the activation energies in the triplet and singlet B_1 states are 28 kcal/mole and 11 kcal/mole, respectively (Fig. 26). These values are evidently the upper limits of the real activation energies, since no minimization has been carried out on the excited structures. They are nevertheless of the same order of magnitude as the excess of energy with which the triplet state is experimentally created, in the photolysis of either diazomethane (28 kcal/mole) or ketene (19 kcal/mole) [256]. The correlation diagram and the "ab-initio" calculations therefore give a partial answer as to why the singlet methylene does not abstract hydrogen atoms. It is because, in the least-motion reaction, this first singlet goes up to a high energy zwitterionic state. The ground-state triplet is a much better vector for this reaction.

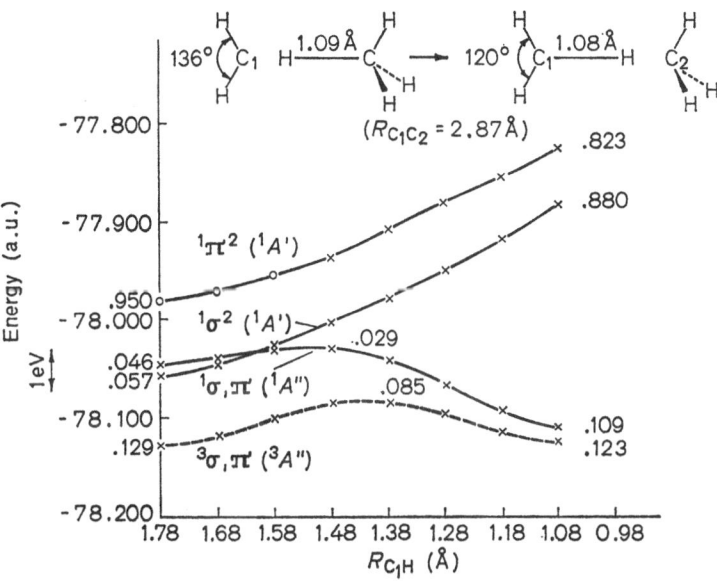

Fig. 26. Energy surfaces for the hydrogen abstraction reaction in the system methylene—methane

3. *The insertion of methylene* into CH paraffinic bonds [256,258] has also been considered by Dewar et al. [239]. There is a valley in the first pathway they consider (a), but even along this valley the energy rises

$$\begin{array}{ccc} \text{H}\!\!>\!\!\text{C} \longrightarrow & \text{H}\!\!>\!\!\text{C}\!\!<\!\!\text{H} & \text{H}\!\!>\!\!\text{C} \longrightarrow \text{H}\!-\!\text{C}\!\!<\!\!\text{H} \\ \text{(a)} & & \text{(b)} \end{array}$$

steadily as the methylene approaches. Neither abstraction nor insertion takes place. The alternative pathway (b) leads, however, to abstraction with the formation of two methyl radicals (in the MINDO/2 treatment the corresponding activation energy is 3.8 kcal/mole). The products formally derived by insertion in fact arise by combination of the pair of radicals formed initially by abstraction.

There is another feature of methylene insertion for which a complete explanation has not yet been proposed. This is the generally observed failure of CH_2 to insert in a CC single bond. Examination of the ab-initio ethane wave function [259] leads to the conclusion that CH_2 insertion is forbidden by electron repulsion for all directions of approach

61

to the CC bond. This is in sharp contrast to insertion into a CH bond, where electron repulsion around the H atom is lowered, and to ylide formation, where there exist paths for avoiding excess repulsion.

Conclusion

In this second part "ab-initio" calculations have been used to investigate various types of photoreactions. These calculations provide quantitative support for the first attempt at a unifying description of photochemical reactivity. Salem's investigation reveals two families of processes.

1. In the hydrogen abstraction by ketones and in the n addition of ketones to electron-rich olefins *two* radical centers are created. In the hydrogen abstraction, for example, σ and π centers appear, respectively, at the carbon atom which the hydrogen is leaving, and in the π system of the carbonyl fragment. Such reactions are dubbed σ,π *bitopic* reactions [194]. More generally, if the two radical centers are labelled A and B, the low-lying electronic states of the products consist of the diradical state D'' (double prime because of the A'' symmetry of this state) $\sigma_A \pi_B$ and of the two zwitterionic states $Z_1'(\sigma_A^2)$ and $Z_2'(\pi_B^2)$ (prime because of the A' symmetry). The general correlation diagram for $\sigma\pi$ bitopic reactions is shown in Fig. 27 (a), where it is assumed that the lowest pair — singlet and triplet — of the reactant states has A'' symmetry, as do $n\pi^*$ states. The second excited singlet state, probably $\pi\pi^*$, has A' symmetry and correlates with the second zwitterionic state Z_2'. $\sigma\pi$ bitopic reactions are characterized by a crossing between the symmetric A' singlet state and the antisymmetric singlet and triplet A''. This region is well suited for efficient internal conversion $(^1A'' \rightarrow {}^1A')$, and even more for intersystem crossing $(^3A'' \rightarrow {}^1A')$, since large spin-orbit coupling elements are expected when the odd electron must switch from a σ-type orbital to a π-type orbital [204,205,261] as is the case here. If one of these two processes takes place, the reacting system returns to the ground-state surface and back to the reactants. $\sigma\pi$ bitopic reactions are facile if they follow throughout the singlet $n\pi^*$ surface and finally give the singlet D'' diradical. In the parent triplet state the energy evolution is equally favorable, but at the end of the pathway intersystem crossing is required from the triplet diradical to the final singlet D''.

More generally, four families of bitopic reactions may be expected. $\sigma\pi$ $(\pi\sigma)$ "odd" bitopic reactions only have been dealt with here, but "even" $(\pi\pi$ and $\sigma\sigma)$ processes are likely to occur (Table 6). In these all the final states are symmetric. Therefore only avoided crossings will be found, but the potential energy curves will still have the memory, and present the characteristics, of their intended crossings.

Table 6. Product states for different possible coplanar bitopic and tritopic reactions

Reaction Type		Diradical States	Zwitterionic States
σ,π	Bitopic (Odd)	1 (Antisymmetric)	2 (all Symmetric)
$\left.\begin{array}{l}\sigma,\sigma\\ \pi,\pi\end{array}\right\}$	Bitopic (Even)	1 (Symmetric)	2 (all Symmetric)
$\left.\begin{array}{l}\sigma,\sigma,\pi\\ \sigma,\pi,\pi\end{array}\right\}$	Tritopic (Odd)	3 $\left\{\begin{array}{l}\text{1 Symmetric}\\ \text{2 Antisymmetric}\end{array}\right.$	3 (all Symmetric)
$\left.\begin{array}{l}\sigma,\sigma,\sigma\\ \pi,\pi,\pi\end{array}\right\}$	Tritopic (Even)	3 (all Symmetric)	3 (all Symmetric)

Taken from Ref. [194]).

2. In the α cleavage of ketones and in the ring-opening of azirines three radical centers are created upon excitation. Two of them (σ_A, σ_B) result from the cleavage of a single AB bond and have σ character, while the third one (π_{BCD}) is delocalized in the π system. Such reactions are christened σ,σ,π *tritopic* reactions. Since each of these three centers may be occupied by 0,1,2 electrons, the manifold of electronic states is more complicated than in the preceding case. First, there are three pairs — singlet and triplet — of diradical states, $\sigma_A\sigma_B$, $\sigma_A\pi_{BCD}$ and $\sigma_B\pi_{BCD}$. The last one, where σ and π electrons are partially localized on the same atom B, is highly energetic and may be ignored. Two radical structures have then to be considered: the lower $\sigma\pi$ diradical D'' (A'' symmetry) and the $\sigma\sigma$ diradical D' (A' symmetry). Each of them gives rise to a singlet and a triplet state. Second, there are three zwitterionic states of A' symmetry (σ_A^2, σ_B^2, π_{BCD}^2). Only the lowest one Z' is included in the correlation diagram. Two situations may occur, depending on the relative position of the two diradicals D' and D'':

a) Fig. 27 (b) schematizes the case where D' is below D'' (as in the cleavage of saturated ketones). The first triplet $n\pi^*$ behaves like the

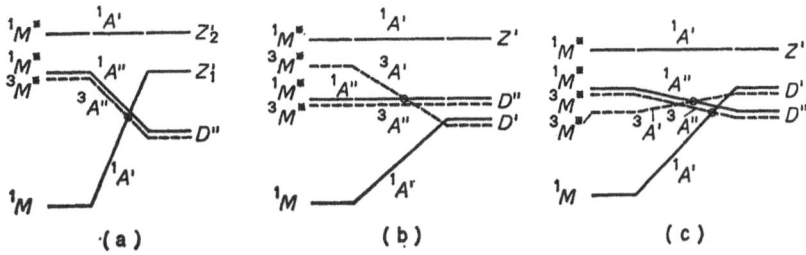

Fig. 27. General state correlation diagram for σ,π bitopic (*a*) and σ,σ,π tritopic (*b c*) *coplanar* reactions

parent $n\pi^*$ singlet, though at a slightly lower energy. The second $\pi\pi^*$ triplet of A' symmetry correlates with D', whereas the still higher parent singlet correlates with Z'. The essential feature of this correlation diagram is the existence of a region where $^3A'$ is crossing both singlet and triplet A'' states. The diradical D'' is probably best obtained directly from the $n\pi^*$ singlet state. D' on the other hand may be obtained directly from the $^3\pi\pi^*$ state or from the singlet $n\pi^*$ with efficient intersystem crossing in the circled region.

b) Fig. 27(c) represents the second case where D'' is below D' (as in the cleavage of dienones). Here again $^3\pi\pi^*$ (now the lowest triplet) intersects the pair of $n\pi^*$ states (but in the opposite sense). Further along the reaction coordinate these $n\pi^*$ states also cross the $^1A'$ ground state but intersystem crossing $(^3A'' \rightarrow {}^1A')$ in this region tends to bring the excited species back to the ground state of the reactants. The diradical D'' seems to be best obtained, in the singlet manifold, directly from the lowest $n\pi^*$ singlet and, in the triplet manifold, directly from the $n\pi^*$ triplet state, or, with a unique internal conversion process, from the $^3\pi\pi^*$ state.

As shown in Table 6, three other families of tritopic reactions — odd $\sigma\pi\pi$ and even $\sigma\sigma\sigma$ and $\pi\pi\pi$ — may also be expected to occur.

To summarize these findings, Salem has suggested that in *coplanar* $\sigma\pi$ bitopic or $\sigma\sigma\pi$ tritopic reactions the antisymmetric diradical D'' originates preferentially from the singlet or triplet $n\pi^*$ states, while the symmetric one D' comes from the $\pi\pi^*$ triplet state. These paths are "direct" in the sense that they do not require any radiationless transition. Of course, in certain cases, the system may prefer a lower energy pathway involving a single radiationless process (as, for example, when the $^{1,3}n\pi^*$ excited azirine reaches the symmetric diradical D'). In *noncoplanar* reactions, where the states are no longer symmetry-differentiated, the crossings between two singlets or triplets become strongly avoided (Fig. 28) and the first singlet state cannot reach the diradical ground state

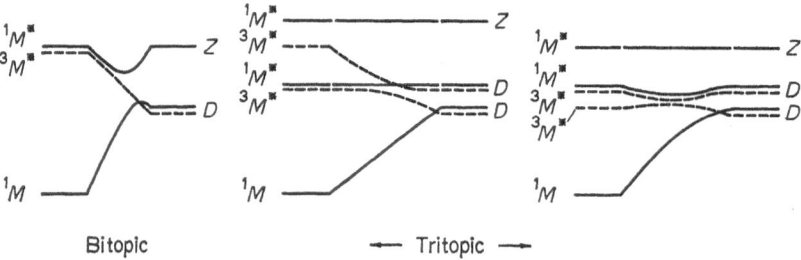

Fig. 28. General state correlation diagram for σ,π bitopic and σ,σ,π tritopic *non-coplanar* reactions

of the products. This ground diradical product originates preferentially in the lowest triplet, whatever its orbital nature (nπ^* in the cleavage of saturated ketones or $\pi\pi^*$ in the cleavage of unsaturated dienonic compounds).

References

[1] Katriel, J.: Theoret. Chim. Acta 23, 309 (1972).

[2] Katriel, J.: Theoret. Chim. Acta 26, 163 (1972).

[3] Davidson, E. R.: J. Chem. Phys. 42, 4199 (1965).

[4] Lemberger, A., Pauncz, R.: Acta Phys. Hung. 27, 169 (1969).

[5] Messmer, R. P., Birss, F. W.: J. Phys. Chem. 73, 2085 (1969).

[6] Williams, J E, Stand, P J., Schleyer, P. v. R.: Ann. Rev. Phys. Chem. 19, 531 (1969).

[7] Lifson, S., Warshel, A.: J. Chem. Phys. 49, 5116 (1968).

[8] Warshel, A., Lifson, S.: J. Chem. Phys. 53, 8582 (1970).

[9] Lewitt, M., Lifson, S.: J. Mol. Biol. 46, 269 (1969).

[10] Lewitt, M.: Nature 224, 759 (1969).

[11] Tric, C.: J. Chem. Phys. 51, 4778 (1969).

[12] Dewar, M. J. S.: The molecular orbital theory of organic chemistry. New York: McGraw Hill 1966.

[13] Salem, L.: The molecular orbital theory of conjugated systems. New York: W. A. Benjamin Inc. 1966.

[14] Daudel, R., Sandorfy, C : Semi-empirical wave mechanical calculations of polyatomic molecules New Haven, Conn : Yale University Press 1971.

[15] Klopman, G , O'Leary, B : Top. Curr. Chem. 15, 415 (1970).

[16] Murrell, J. N., Harget, A. J.: "Semi-empirical SCF molecular orbital theory of molecules. New York: John Wiley and Sons 1972.

[17] Hoffmann, R., Swaminathan, S., Odell, B. G., Gleiter, R.: J. Am. Chem. Soc. 92, 7091 (1970).

[18] Becker, R. S., Inuzuka, K., Balke, D. E.: J. Am. Chem. Soc. 93, 38 (1971).

[19] Becker, R. S., Inuzuka, K., King, J., Balke, D. E.: J. Am. Chem. Soc. 93, 43 (1971).

[20] Becker, R. S., Inuzuka, K., King, J.: J. Chem. Phys. 52, 5164 (1970).

[21] Hoffmann, R.: IUPAC Photochemistry 3 (St Moritz, 1970), p. 578. London: Butterworths.

[22] Zimmermann, H. E., Binkley, R. W., Givens, R. S., Grunewald, G. L., Sherwin, M. A.: J. Am. Chem. Soc. 91, 3316 (1969).

[23] Segal, G. A.: J. Chem. Phys. 53, 360 (1970).

[24] White, G. M., Yarwood, A. J., Santry, D. P.: Chem. Phys. Letters 13, 501 (1972).

[25] Dixon, R. N.: Mol. Phys. 12, 83 (1967).

[26] Dewar, M. J. S., Haselbach, E.: J. Am. Chem. Soc. 92, 590 (1970).

[27] Bodor, N., Dewar, M. J. S., Harget, A., Haselbach, E.: J. Am. Chem. Soc. 92, 3584 (1970).

[28] Warshel, A., Karplus, M.: J. Am. Chem. Soc. 94, 5612 (1972).

[29] Kelly, H. P.: Phys. Rev. 136B, 896 (1964).

[30] Whitten, J. L., Hackmeyer, M.: J. Chem. Phys. 51, 5584 (1969).

[31] Jorgensen, W. L., Salem, L.: The organic chemist's book of orbitals, p. 84. New York: Academic Press 1973.

[32] Herzberg, G.: Electronic spectra and electronic structure of polyatomic molecules. Princeton, N. J.: D. Van Nostrand Co., Inc. 1967.
[33] Coon, J. B., Naugle, N. W., McKenzie, R. D.: J. Mol. Spectry. 20, 107 (1966).
[34] Job, V. A., Sethuraman, V., Innes, K. K.: J. Mol. Spectry. 30, 365 (1969).
[35] Jones, V. T., Coon, J. B.: J. Mol. Spectry. 31, 137 (1969).
[36] Callomon, J. H., Innes, K. K.: J. Mol. Spectry 10, 166 (1963).
[37] Brand, J. C. D.: J. Chem. Soc. 1956, 858.
[38] Robinson, G. W.: Can. J. Phys. 34, 699 (1956).
[39] Di Giorgio, V. E., Robinson, G. W.: J. Chem. Phys. 31, 1678 (1959).
[40] Oka, T.: J. Phys. Soc. Japan 15, 2274 (1960).
[41] Takagi, K., Oka, T.: J. Phys. Soc. Japan 18, 1174 (1963).
[42] Roothaan, C. C. J.: Rev. Mod. Phys. 23, 69 (1951).
[43] Wannier, G. H.: Phys. Rev. 52, 191 (1937).
[44] Lennard-Jones, J.: Proc. Roy. Soc. (London) A 198, 1, 14 (1949).
[45] Pople, J. A.: Quart. Rev. (London) 11, 273 (1973).
[46] Edmiston, C., Ruedenberg, K.: Rev. Mod. Phys. 35, 457 (1963).
[47] Goddard III, W. A.: Phys. Rev. 157, 73, 81 (1967).
[48] Kolos, W., Roothaan, C. C. J.: Rev. Mod. Phys. 32, 219 (1960).
[49] Roothaan, C. C. J., Sachs, L. M., Weiss A. W.: Rev. Mod. Phys. 32, 186 (1960).
[50] Cade, P. E., Sales, K. D., Wahl, A. C.: J. Chem. Phys. 44, 1973 (1966).
[51] Kahalas, S. L., Nesbet, R. K.: J. Chem. Phys. 39, 529 (1963).
[52] Löwdin, P. O.: Advan. Chem. Phys. 2, 207 (1959).
[53] Buenker, R. J., Peyerimhoff, S. D.: J. Chem. Phys. 53, 1368 (1970).
[54] Wahl, A. C., Bertoncini, P. J.: Phys. Rev. Letters 25, 991 (1970).
[55] Das, G.: J. Chem. Phys. 46, 1568 (1967).
[56] Veillard, A., Clementi, E.: Theoret. Chim. Acta 7, 133 (1967).
[57] Hinze, J., Roothaan, C. C. J.: Progr. Theoret. Phys. (Kyoto) 40, 37 (1967).
[58] Sabelli, N., Hinze, J.: J. Chem. Phys. 50, 684 (1969).
[59] Yutsis, A. P.: Zh. Eksperim. Teor. Fiz. 23, 129 (1952).
[60] Buenker, R. J., Whitten, J. L., Petke, J. D.: J. Chem. Phys. 49, 2261 (1968).
[61] Morokuma, K., Konishi, H.: J. Chem. Phys. 55, 402 (1971).
[62] Hayes, D. M., Morokuma, K.: Chem. Phys. Letters 12, 539 (1972).
[63] Hehre, W. J., Stewart, R. F., Pople, J. A.: J. Chem. Phys. 51, 2657 (1970).
[64] Huzinaga, S., Arnau, C.: J. Chem. Phys. 54, 1948 (1971).
[65] Huzinaga, S., Arnau, C.: Phys. Rev. A 1, 1285 (1970).
[66] Hunt, W. J., Goddard III, W. A.: Chem. Phys. Letters 3, 414 (1969).
[67] Lefebvre-Brion, H., Moser, C., Nesbet, R.: J. Mol. Spectry. 13, 418 (1964).
[68] Dunning, T. H., Hunt, W. J., Goddard III, W. A.: Chem. Phys. Letters 4, 147 (1969).
[69] Wilkinson, P. G., Johnston, W. L.: J. Chem. Phys. 18, 190 (1950).
[70] McWilliams, D., Huzinaga, S.: J. Chem. Phys. 55, 2604 (1971).
[71] Morokuma, K., Iwata, S.: Chem. Phys. Letters 16, 192 (1972).
[72] Iwata, S., Morokuma, K.: Chem. Phys. Letters 19, 94 (1973).
[73] Hartree, D. R., Proc. Roy. Soc. (London) A 154, 588 (1936).
[74] Slater, J. C.: Phys. Rev. 82, 538 (1951).
[75] Löwdin, P. O.: Phys. Rev. 97, 1474 (1955).
[76] Pople, J. A., Nesbet, R. K.: J. Chem. Soc. 1954, 571.
[77] Berthier, G., Compt. Rend. 238, 91 (1954).
[78] Berthier, G.: J. Chim. Phys. 51, 137, 363 (1954).
[79] Berthier, G.: In: Molecular orbitals in chemistry, physics and biology, p. 57. New York: Academic Press 1964.
[80] Nesbet, R. K.: Proc. Roy. Soc. (London) A 230, 312, 322 (1955).

81) Roothaan, C. C. J.: Rev. Mod. Phys. *32*, 179 (1960).
82) Roothaan, C. C. J., Bagus, P. S.: Methods Comput. Phys. *2*, 47 (1963).
83) Nesbet, R. K.: Rev. Mod. Phys. *33*, 28 (1961).
84) Nesbet, R. K.: Rev. Mod. Phys. *35*, 552 (1963).
85) Basch, H., Robin, M. A., Kuebler, N. A.: J. Chem. Phys. *49*, 5007 (1968).
86) Buenker, R. J., Peyerimhoff, S. D.: Theoret. Chim. Acta *12*, 183 (1968).
87) Salem, L.: Bull. Soc. Chim. France 1970, 3161.
88) Jean, Y., Salem, L.: Chem. Commun. *1971*, 322.
89) Jean, Y., Salem, L., Wright, J. S., Horsley, J. A., Moser, C., Stevens, R. M.: XXIII IUPAC Meeting (Boston 1971), p. 197. London: Butterworths.
90) Fink, W. H.: J. Am. Chem. Soc. *94*, 1073, 1078 (1972).
91) Salem, L., Leforestier, C., Segal, G., Wetmore, R.: to be published.
92) Löwdin, P. O.: Advan. Chem. Phys. *2*, 207 (1959).
93) Nesbet, R. K.: Advan. Chem. Phys. *9*, 321 (1966).
94) Kutzelnigg, W.: In: Selected topics in molecular physics, p. 91. Weinheim: Verlag Chemie 1972.
95) Michl, J.: Top. Curr. Chem. *46*, 1 (1974).
96) Zimmerman, H. E., Acc. Chem. Res. *4*, 272 (1971).
97) Van der Lugt, W. Th. A. M., Oosterhoff, L. J.: J. Am. Chem. Soc. *91*, 6042 (1969).
98) Dougherty, R. C.: J. Am. Chem. Soc. *93*, 7187 (1971).
99) Woodward, R. B., Hoffmann, R.: Angew. Chem. Intern. Ed. Engl. *8*, 781 (1969).
100) Fukui, K.: Acc. Chem. Res. *4*, 57 (1971).
101) Fukui, K., Top. Curr. Chem. *15*, 1 (1970).
102) Longuet-Higgins, H. C., Abrahamson, E. W.: J. Am. Chem. Soc. *87*, 2045 (1965).
103) Bryce-Smith, D.: Chem. Commun. *1969*, 806.
104) Michl, J.: J. Am. Chem. Soc. *93*, 523 (1971).
105) Michl, J.: Mol. Photochem. *4*, 287 (1972).
106) Chu, N. Y. C., Kearns, D. R.: J. Phys. Chem. *74*, 1255 (1970).
107) Epiotis, N. D.: J. Am. Chem. Soc. *94*, 1941, 1946 (1972).
108) Constanciel, R.: Theoret. Chim. Acta *26*, 249 (1972).
109) Godfrey, M.: J. Chem. Soc. Perkin II *1972*, 1690.
110) Herndon, W. C., Giles, W. B.: Mol. Photochem. *2*, 277 (1970).
111) Herndon, W. C.: Top. Curr. Chem. to be published.
112) Pearson, R. G.: Chem. Phys. Letters *10*, 31 (1971).
113) Devaquet, A.: J. Am. Chem. Soc. *94*, 5626, 9012 (1972).
114) Zemke, W. T., Lykos, P. G., Wahl, A. C.: J. Chem. Phys. *51*, 5635 (1969).
115) Brändas, E.: Intern. J. Quant. Chem. *1*, 847 (1967).
116) Brändas, E.: Intern. J. Quant. Chem. *2*, 391 (1968).
117) Jug, K., Lykos, P. G., McLean, A. D.: Theoret. Chim. Acta *25*, 17, 41 (1972).
118) Gupta, B. K., Matsen, F. A.: J. Chem. Phys. *50*, 3797 (1969).
119) Schaeffer III, H. F., Miller, W. H.: J. Chem. Phys. *55*, 4107 (1971).
120) Schaeffer III, H. F.: J. Chem. Phys. *54*, 2207 (1971).
121) Kutzelnigg, W., Staemmler, V., Gelus, M.: Chem. Phys. Letters *13*, 496 (1972).
122) Melius, C. F., Goddard III, W. A.: J. Chem. Phys. *56*, 3348 (1972).
123) Docken, K. K., Hinze, J.: J. Chem. Phys. *57*, 4928 (1972).
124) Bertoncini, P. J., Das, G., Wahl, A. C.: J. Chem. Phys. *52*, 5112 (1970).
125) Heil, T. G., Schaeffer III, H. F.: J. Chem. Phys. *56*, 958 (1972).
126) Huron, B., Rancurel, P.: Chem. Phys. Letters *13*, 515 (1972).
127) Rubinstein, M., Shavitt, I.: J. Chem. Phys. *51*, 2014 (1969).
128) Hay, P. J., Goddard III, W. A.: Chem. Phys. Letters *14*, 46 (1972).

129) Devaquet, A., Ryan, J. A.: Chem. Phys. Letters 22, 269 (1973).
130) Grimbert, D., Devaquet, A.: Mol. Phys., 27, 831 (1974).
131) Hay, P. J., Hunt, W. J., Goddard III, W. A.: J. Am. Chem. Soc. 94, 638 (1972).
132) Allwood, P. J., Kazmaier, P. M., Rauk, A.: J. Am. Chem. Soc. 95, 5466 (1973).
133) Ditchfield, R., Del Bene, J. E., Pople, J. A.: J. Am. Chem. Soc. 94, 4806 (1972).
134) Del Bene, J. E.: J. Am. Chem. Soc. 94, 3713 (1972).
135) Mulliken, R. S.: Phys. Rev. 43, 279 (1933).
136) Mulliken, R. S., Roothaan, C. C. J.: Chem. Rev. 41, 219 (1947).
137) Baird, N. C., West, R. M.: J. Am. Chem. Soc. 93, 4427 (1971).
138) Kaldor, U., Shavitt, I.: J. Chem. Phys. 48, 191 (1968).
139) Evans, D. F.: J. Chem. Soc. 1960, 1735.
140) Moskowitz, J. W., Harrison, M. C.: J. Chem. Phys. 42, 1726 (1965).
141) Schulman, J. M., Moskowitz, J. W., Hollister, C.: J. Chem. Phys. 46, 2759 (1965).
142) Rabinovitch, B. S., Douglas, J. E., Looney, F. S.: J. Chem. Phys. 20, 1807 (1952).
143) Magee, J. L., Shand, W., Eyring, H.: J. Am. Chem. Soc. 63, 677 (1941).
144) Harman, R. A., Eyring, H.: J. Chem. Phys. 10, 557 (1942).
145) Kistiakowsky, G. B., Smith, W. R.: J. Am. Chem. Soc. 58, 766 (1936).
146) Walsh, A. D.: J. Chem. Soc. 1953, 2325.
147) Lorquet, A. J.: J. Phys. Chem. 74, 895 (1970).
148) Burnelle, L., Litt, C.: Mol. Phys. 9, 433 (1955).
149) Burnelle, L.: J. Chem. Phys. 43, S 29 (1965).
150) Baird, N. C., Swenson, J. R.: Chem. Phys. Letters 22, 183 (1973).
151) Lathan, W. A., Hehre, W. J., Curtiss, L. A., Pople, J. A.: J. Am. Chem. Soc. 93, 6377 (1972).
152) Devaquet, A., Salem, L.: Can. J. Chem. 49, 977 (1971).
153) Devaquet, A.: J. Am. Chem. Soc. 94, 5160 (1972).
154) Baird, N. C., West, R. M.: Mol. Photochem. 5, 209 (1973).
155) Guzzo, A. V., Pool, G. L.: J. Phys. Chem. 73, 2512 (1969).
156) Solly, R. K., Golden, D. M., Benson, S. W.: Intern. J. Chem. Kinetics 2, 11, 381 (1970).
157) Golden, D. M., Benson, S. W.: Chem. Rev. 69, 125 (1969).
158) Hollas, J. M.: Spectrochim. Acta 19, 1425 (1963).
159) Brand, J. C. D., Williamson, D. G.: Discussions Faraday Soc. 35, 184 (1962).
160) Inuzuka, K.: Bull. Chem. Soc. Japan 34, 729 (1961).
161) Bair, E. J., Goetz, W., Ramsay, D. A.: Can. J. Phys. 49, 2710 (1971).
162) Hehre, W. J., Devaquet, A.: to be published.
163) Dewar, M. J. S., Harget, A. J.: Proc. Roy. Soc. (London). Ser. A 315, 433 (1970).
164) Snyder, L. E., Buhl, D., Zuckerman, B., Palmer, P.: Phys. Rev. Letters 22, 679 (1969).
165) De Graff, B. A., Calvert, J. G.: J. Am. Chem. Soc. 89, 2247 (1967).
166) Glicker, S., Stief, L. J.: J. Chem. Phys. 54, 2852 (1971).
167) Abrahamson, E. W., Littler, J. G. F., Vo, K. P.: J. Chem. Phys. 44, 4082 (1966).
168) Walsh, R., Benson, S. W.: J. Am. Chem. Soc. 88, 4570 (1966).
169) McQuigg, R. D., Calvert, J. G.: J. Am. Chem. Soc. 91, 1590 (1969).
170) Gorin, E.: J. Chem. Phys. 7, 256 (1939).
171) Klein, R., Schoen, L. J.: J. Chem. Phys. 24, 1094 (1956).
172) Winter, N. W., Dunning, T. H., Letcher, J. H.: J. Chem. Phys. 49, 1871 (1968).
173) Wagner, P. J., Hammond, G. S.: Advan. Photochem. 5, 89 (1969).

174) Lee, E. K. C.: J. Phys. Chem. *71*, 2804 (1967).
175) Denschlag, H. O., Lee, E. K. C.: J. Am. Chem. Soc. *90*, 3628 (1968).
176) Shortridge, Jr., J. G., Lee, E. K. C.: J. Am. Chem. Soc. *92*, 2228 (1970).
177) Shortridge, Jr., J. G., Rusbult, C. F., Lee, E. K. C.: J. Am. Chem. Soc. *93*, 1863 (1971).
178) Solomon, J., Jonah, C., Chandra, P., Bersohn, R.: J. Chem. Phys. *55*, 1908 (1971).
179) Breuer, G. M., Lee, E. K. C.: J. Phys. Chem. *75*, 989 (1969).
180) Halpern, A. M., Ware, W. R.: J. Chem. Phys. *53*, 1969 (1970).
181) Schlag, E. W., von Weyssenhoff, H.: J. Chem. Phys. *51*, 2508 (1969).
182) Yang, N. C., Feit, E. D., IIui, M. II., Turro, N. J., Dalton, J. C.: J. Am. Chem. Soc. *92*, 6974 (1970).
183) Dalton, J. C., Pond, D. M., Weiss, D. S., Lewis, F. D., Turro, N. J.: J. Am. Chem. Soc. *92*, 2564 (1972).
184) Yang, N. C., Feit, E. D.: J. Am. Chem. Soc. *90*, 504 (1968).
185) Dunion, P., Trumbore, C. N.: J. Am. Chem. Soc. *87*, 4211 (1965).
186) Wagner, P. J., Spoerke, R. W.: J. Am. Chem. Soc. *91*, 4437 (1969).
187) Simonaïtis, R., Cowell, G. W., Pitts, Jr., J. N.: Tetrahedron Letters *1957*, 3751.
188) Norrish, R. G. W.: Trans. Faraday Soc. *33*, 1521 (1937).
189) Kraus, J. W., Calvert, J. G.: J. Am. Chem. Soc. *79*, 5921 (1957).
190) Srinivasan, R.: Advan. Photochem. *1*, 83 (1963).
191) Carless, H. A. J., Metcalfe, J., Lee, E. K. C.: J. Am. Chem. Soc. *94*, 7221 (1972).
192) Wagner, P. J., Kelso, F. A., Zepp, K. G.: J. Am. Chem. Soc. *94*, 7450 (1972).
193) Dalton, J. C., Dawes, K., Turro, N. J., Weiss, D. S., Barltrop, J. A., Coyle, J. D.: J. Am. Chem. Soc. *93*, 7213 (1971).
194) Salem, L.: J. Am. Chem. Soc. *96*, 3486 (1974).
195) Salem, L., Dauben, W. G., Turro, N. J.: J. Chim. Phys. *70*, 694 (1973).
196) Johns, J. W. C., Priddle, S. H., Ramsay, D. A.: Discussions Faraday Soc. *35*, 90 (1963).
197) Adrian, F. J., Cochran, E. L., Bowers, V. A.: J. Chem. Phys. *36*, 1661 (1962).
198) Roncin, J.: private communication to L. Salem.
199) Quinkert, G.: Pure Appl. Chem. *35*, 285 (1973).
200) Padwa, A., Dharan, M., Smolanoff, J., Wetmore, S. I.: Pure Appl. Chem. *33*, 269 (1973).
201) Padwa, A., Dharan, M., Smolanoff, J., Wetmore, S. I.: J. Am. Chem. Soc. *95*, 1945, 1954 (1973).
202) Claus, P., Doppler, Th., Gakis, N., Georgarakis, M., Giezendanner, H., Gilgen, P., Heingartner, H., Jackson, B., Märky, M., Narasiman, N. S., Rosenkranz, H. J., Wunderli, A., Hansen, H. J., Schmid, H.: Pure Appl. Chem. *33*, 339 (1973).
203) Singh, B., Zweig, A., Gallivan, J. B.: J. Am. Chem. Soc. *94*, 1199 (1972).
204) Salem, L., Rowland, C.: Angew. Chem. Intern. Ed. Engl. *11*, 92 (1972).
205) Salem, L.: Pure Appl. Chem. *33*, 317 (1973).
206) Yang, N. C., Yang, D. H.: J. Am. Chem. Soc. *80*, 2913 (1958).
207) Wagner, P. J.: Acc. Chem. Res. *4*, 168 (1971).
208) Turro, N. J., Dalton, J. C., Dawes, K., Farrington, G., Hautala, R., Morton, D., Niemczyk, M., Schore, N.: Acc. Chem. Res. *5*, 92 (1972).
209) Coyle, J. D., Carless, H. A. J.: Chem. Soc. Rev. *11*, 465 (1972).
210) Casey, C. P., Boggs, R. A.: J. Am. Chem. Soc. *94*, 6457 (1972).
211) Garnier, R. A., Schreiber, W. L., Agosta, W. C.: Chem. Commun. *1972*, 729.
212) Michaels, J. L., Noyes, Jr., W. A.: J. Am. Chem. Soc. *85*, 1027 (1963).

213) Ausloos, P., Rebbert, R. B.: J. Am. Chem. Soc. *86*, 4512 (1964).
214) Wagner, P. J., Hammond, G.: J. Am. Chem. Soc. *87*, 4010 (1965).
215) Dougherty, T.: J. Am. Chem. Soc. *87*, 4011 (1965).
216) Coulson, D. R., Yang, N. C.: J. Am. Chem. Soc. *88*, 4511 (1966).
217) Evleth, E. M.: Chem. Phys. Letters *3*, 22 (1969).
218) Wilhite, D. L., Whitten, J. L.: J. Am. Chem. Soc. *93*, 2858 (1971).
219) Wagner, P. J., Kelso, P. A., Kempainnen, A. E., Zepp, R. G.: J. Am. Chem. Soc. *94*, 7500 (1972).
220) Lewis, F. D.: Mol. Photochem. *4*, 501 (1972).
221) Paterno, E., Chieffi, G.: Gazz. Chim. Ital. *39*, 341 (1909).
222) Büchi, G., Inman, C. G., Lipinsky, E. S.: J. Am. Chem. Soc. *76*, 4327 (1954).
223) Büchi, G., Kofron, J. T., Koller, E., Rosenthal, D.: J. Am. Chem. Soc. *78*, 876 (1956).
224) Arnold, D. R.: Advan. Photochem. *6*, 301 (1968).
225) Caldwell, R. A., Sovocool, G. W., Gajewski, R. P.: J. Am. Chem. Soc. *95*, 2549 (1973).
226) Herzberg, G.: Proc. Roy. Soc. (London) *A 262*, 291 (1961).
227) Herzberg, G., Johns, J. W. C.: Proc. Roy. Soc. (London) *A 295*, 107 (1966).
228) Herzberg, G., Johns, J. W. C.: J. Chem. Phys. *54*, 2276 (1971).
229) Wasserman, E., Yager, W. A., Kuck, V. J.: Chem. Phys. Letters *7*, 409 (1970).
230) Wasserman, E., Kuck, V. J., Hutton, R. S., Yager, W. A.: J. Am. Chem. Soc. *92*, 7491 (1970).
231) Halberstadt, M. L., McNesby, J. R.: J. Am. Chem. Soc. *89*, 3417 (1967).
232) Carr, R. W., Eder, T. W., Topor, M. G.: J. Chem. Phys. *53*, 4716 (1970).
233) Braun, W., Bass, A. M., Pilling, M.: J. Chem. Phys. *52*, 5131 (1970).
234) Staemmler, V.: Theoret. Chim. Acta *31*, 49 (1973) and references therein.
235) Foster, J. M., Boys, S. F.: Rev. Mod. Phys. *32*, 300 (1960).
236) Skell, P. S., Woodworth, R. C.: J. Am. Chem. Soc. *78*, 4496 (1956).
237) von E. Doering, W., La Flamme, P.: J. Am. Chem. Soc. *78*, 5447 (1956).
238) Kistiakowsky, G. B., Sauer, K.: J. Am. Chem. Soc. *78*, 5699 (1956).
239) Bodor, N., Dewar, M. J. S., Wasson, J. S.: J. Am. Chem. Soc. *94*, 9095 (1972).
240) Dewar, M. J. S.: Top. Curr. Chem. to be published.
241) Dewar, M. J. S., Trinajstic, N.: J. Chem. Soc. D *1970*, 646.
242) Eder, T. W., Carr, R. W.: J. Phys. Chem. *73*, 2074 (1969).
243) Ho, S., Unger, I., Noyes, Jr., W. A.: J. Am. Chem. Soc. *80*, 5091 (1967).
244) King, D. R., Rabinovitch, B. S.: J. Phys. Chem. *72*, 191 (1968).
245) Carr, R. W., Kistiakowsky, G. B.: J. Phys. Chem. *70*, 118 (1966).
246) McKnight, C., Lee, E. K. C., Rowland, F. S.: Ber. Bunsenges. Physik. Chem. *72*, 236 (1968).
247) Dewar, M. J. S.: Bull. Soc. Chim. France *18*, C71 (1951).
248) Gesser, H., Steacies, E. W. R.: Can. J. Chem. *34*, 113 (1956).
249) Frey, H. M., Kistiakowsky, G. B.: J. Am. Chem. Soc. *79*, 637 (1957).
250) Frey, H. M.: Proc. Roy. Soc. (London) *A 250*, 409 (1959).
251) Richardson, D. B., Simons, M. C., Dvoretzky, I.: J. Am. Chem. Soc. *82*, 5001 (1960).
252) Richardson, D. B., Simons, M. C., Dvoretzky, I.: J. Am. Chem. Soc. *83*, 1934 (1961).
253) Frey, H. M., Walsh, R.: J. Chem. Soc. *A 2115* (1971).
254) Roth, H. D.: J. Am. Chem. Soc. *94*, 1961 (1972).
255) Roth, H. D.: J. Am. Chem. Soc. *93*, 1527, 4935 (1971).
256) von E. Doering, W., Buttery, R. G., Langhlin, R. G., Chaudhuri, N.: J. Am. Chem. Soc. *78*, 3224 (1956).

257) von E. Doering, W., Prinzbach, H.: Tetrahedron *6*, 24 (1959).
258) von E. Doering, W., Knox, L. H.: J. Am. Chem. Soc. *83*, 1989 (1961).
259) Fink, W. H., Allen, L. C.: J. Chem. Phys. *46*, 2261 (1967).
260) Bender, C. F., Schaffer III, H. F., Franceschetti, D. R., Allen, L. C.: J. Am. Chem. Soc. *94*, **6888** (1972).
261) El Sayed, M. A.: Acc. Chem. Res. *1*, 8 (1968).

Received February 1, 1974

Photochemistry of β, γ -Unsaturated Ketones

Prof. William G. Dauben, Dr. Gerrit Lodder*, and Prof. Junes Ipaktschi**

Department of Chemistry, University of California, Berkeley, California 94720, USA

Contents

Present Address:

* University of Leiden, Holland
** Fachbereich Chemie der Philipps-Universität Marburg/Lahn

The photochemistry of β,γ-unsaturated ketones can be subdivided into the reactions of the singlet and the triplet state. The structural parameters which effect the reactivity of each of these states, and, in turn the efficiency of intersystem crossing have been categorized. The most general reaction displayed by this chromophore is isomerization via a 1,3-acyl shift or via a 1,2-acyl shift (oxa-di-π-methane rearrangement). These reactions can occur from either state but a strong preference for the former to occur via the singlet state and the latter to occur via the triplet state is found. A wide variety of other processes occur and again their efficiencies show a structural dependence. The parameters have been evaluated.

Introduction

As a class of compounds whose photochemistry has been intensively studied, the β,γ-unsaturated ketones are of recent vintage. In particular, the chemistry from the triplet state of these compounds has only received attention during the last few years. All rapidly expanding fields in chemistry show incongruities and inconsistencies but a careful survey of the results published about the photochemistry of this class of nonconjugated ketones shows that this field may have gotten more than its share. It is the purpose of this review to point out the general trends and similarities in the field but at the same time to mention the dissimilarities which, if avoided, oversimplify the complex photochemical personality of β,γ-unsaturated ketones.

Critical evaluation of the available facts has been hampered in a number of cases because systematic studies have not been performed; this feature is especially true with regard to the multiplicity of the state from which a reaction occurs. Therefore, it will be sometimes necessary to rely upon analogy between reactions of very closely related compounds, appreciating the fact that nothing should be more suspect than analogy in a field where, as the reader will see, about every generality has its exception.

β,γ-Unsaturated ketones formally contain two separate chromophoric groupings, the carbonyl group and the alkene group. When these two groupings are not in the same plane they are coupled, the intensity of the nominal n,π^* band of the carbonyl group being 10—100 times larger than that of an isolated carbonyl group.[1] This intensity enhancement of the n,π^* band can be the result either of its mixing with a π_{CC},π^*_{CO} charge transfer band which appears at a shorter wavelength or by a direct mixing with the π_{CC},π^*_{CC} transition.[1–3] Enhanced intensity is also seen in optical rotatory dispersion and circular dichroism measurements.[3]

74

The coupling of these chromophoric groups in cyclopent-3-en-1-one and norborn-2-en-5 (or 7)-one has also been detected by photoelectron spectroscopy. [4]

The triplet state of a β,γ-unsaturated ketone also shows special spectroscopic properties. From phosphorescence studies at 77 °K, [5,6] it has been shown that the lowest triplets have an π,π^* configuration with energies from 68—74 kcal/mol.

The foregoing observations are in line with CNDO/S calculations with configuration interaction for *trans*-4-hexene-2-one [7] which give a lowest excited singlet state which is mainly of n,π* (54%) and alkene-carbonyl charge transfer (32%) configuration, and which show the lowest triplet state consists mainly (70%) of a π,π^*-configuration.

A variety of photochemical reactions have been found for both the singlet and triplet excited states of β,γ-unsaturated ketones. It has been the molecular rearrangements, however, which appear to be the unique character of this grouping, rearrangements which are controlled by the singlet or triplet nature of the excited state.

The most often encountered rearrangement reaction of the singlet excited state is what is formally an 1,3-acyl shift (*1*→*2*) and that of the triplet is what is formally an 1,2-acyl shift (*1*→*3*).

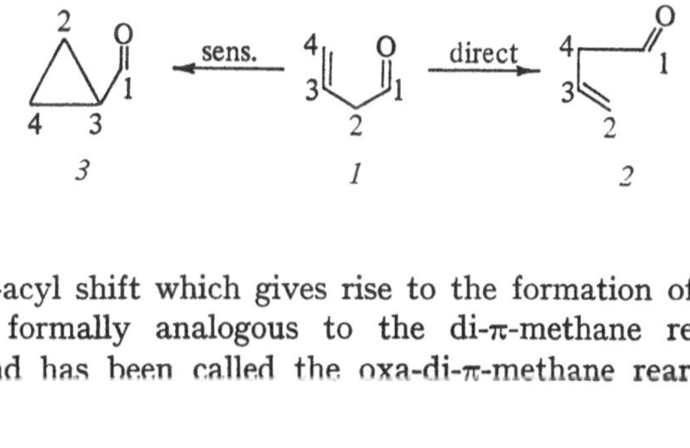

The 1,2-acyl shift which gives rise to the formation of cyclopropyl ketones is formally analogous to the di-π-methane rearrangement (*4*→*5*), [8] and has been called the oxa-di-π-methane rearrangement. [9]

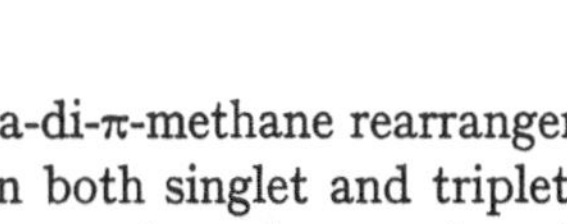

In distinction to the oxa-di-π-methane rearrangement, the di-π-methane rearrangement occurs in both singlet and triplet excited state.

In the Eqs. (1—7), a number of examples of the dichotomy of this singlet versus triplet behavior is listed. From the variety of compounds that give the 1,3-acyl shift from the excited singlet state and the oxa-di-

6

(1) [10,11,12]

10 7 8 9

(2) [13–16]

11

(3) [9]

12

(4) [17]

(5) [15,16]

13 14

(6) [18]

17 15 16

(7) [19,20]

a) X = CH_2 b) X = ▽ c) X = C=C<

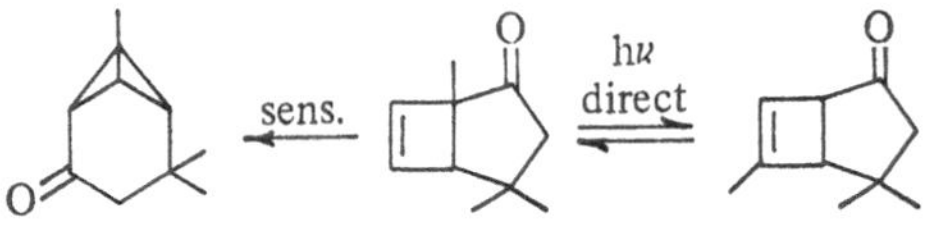

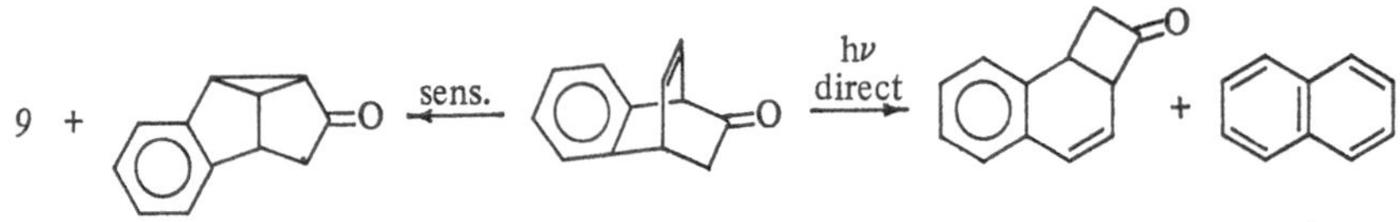

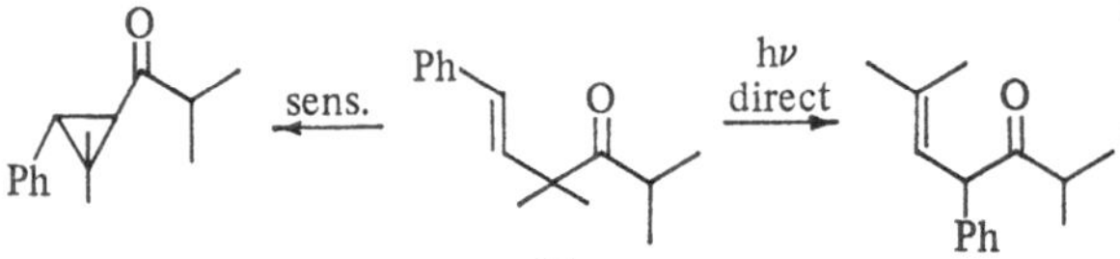

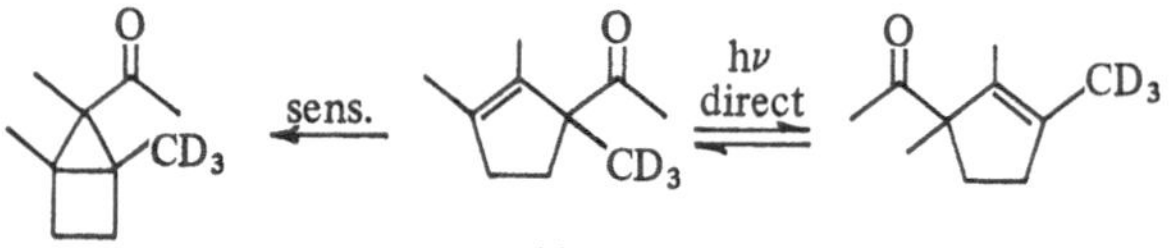

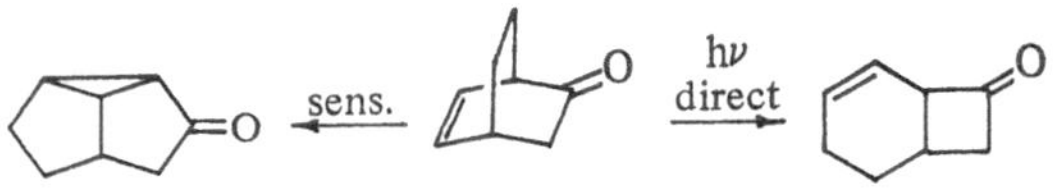

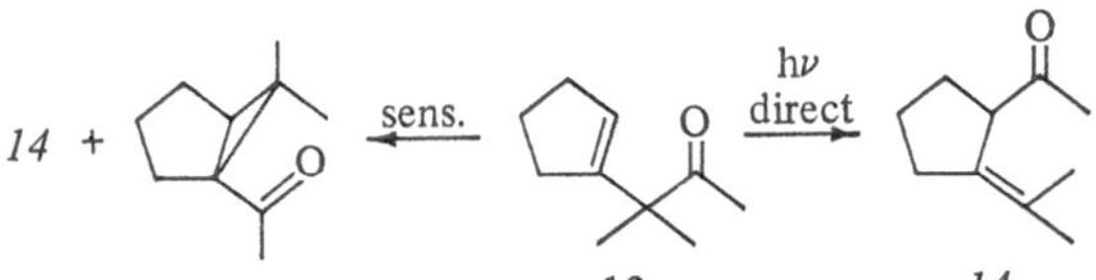

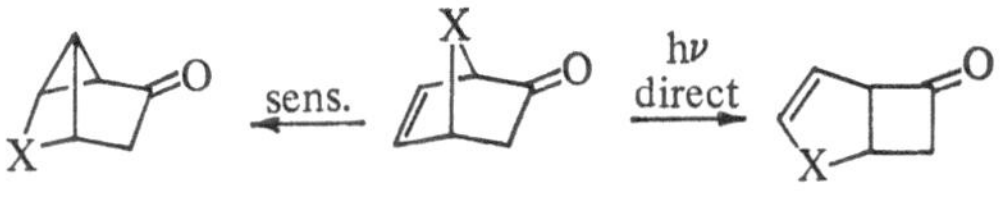

π-methane rearrangement from the triplet state one gathers that these reactions are quite general. From the fact that on excitation of the singlet state no products derived from the triplet state are formed, one concludes that intersystem crossing (*ics*) from the excited singlet to the triplet state is relatively unimportant in these systems.[13,17]

These selected examples follow the often quoted generalities that in β,γ-unsaturated ketones: a) excited singlet-triplet *isc* is inefficient; b) excited singlets give 1,3-acyl shifts and; c) triplets give the oxa-di-π-methane rearrangement. Testing of these generalities against all existing examples of the photobehavior of β,γ-unsaturated ketones shows that they are certainly useful as guidelines but cannot be considered infallible.

From singlet state it has been found that the following reactions often compete with the 1,3-rearrangements: *cis-trans* isomerization, decarbonylation, aldehyde formation, ketene formation, olefin reduction, Norrish type-II cleavage with cyclobutanol formation, [2+2] cycloaddition, and intersystem crossing with concomitant 1,2-acyl shift. From the triplet state a similar series of reactions have been reported including 1,3-acyl shift.

Due to the multitude of reactions from either ecxited state, it is worthwhile to give a systematic survey of the various reactions which are found to occur from the excited singlet and triplet state. By consideration of the many examples, often it is possible to understand, or at least offer a plausible explanation, why certain compounds digress from the above simple generalities. Nevertheless, there are many instances where the results do not readily fit into the framework of our current knowledge.

I. Photoisomerizations

1. 1,3-Acyl Shifts

In the preceding section, Eqs. (1—7) have listed examples of this process and a few more representative examples of this singlet reaction are given in Eqs. (8—12). In a number of these cases the 1,3-shift is a reversible [see Eqs. (1, 4, 8—11)] and in those cases where one of the components is not drained off by a competing reaction, a photoequilibrium is established between the two products. In the few cases where this is found, the equilibrium composition does not always directly reflect the absorption coefficient of the two components of the mixture, indicating that the quantum yield for the forward and reverse reaction may be different. These quantum yields have only been measured in a few cases, [Eqs. (13, 14)], [28,29] and indeed, such has been found.

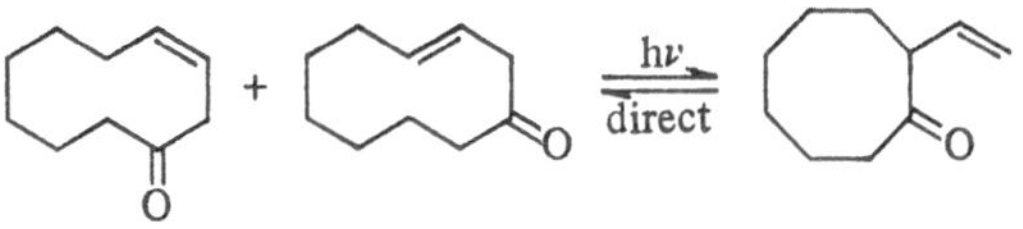

(8) [21]

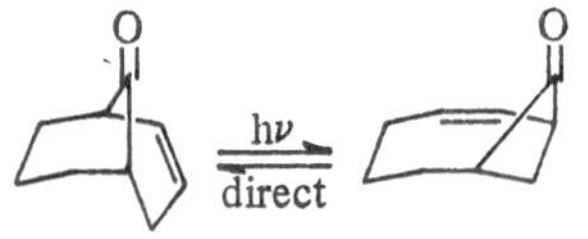

(9) [22,23]

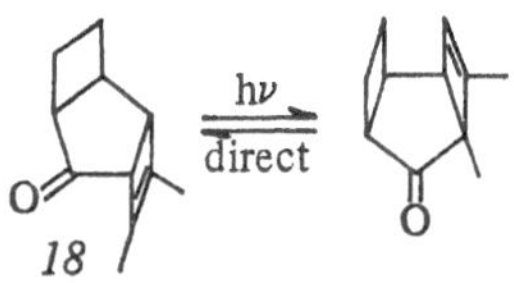

(10) [24]

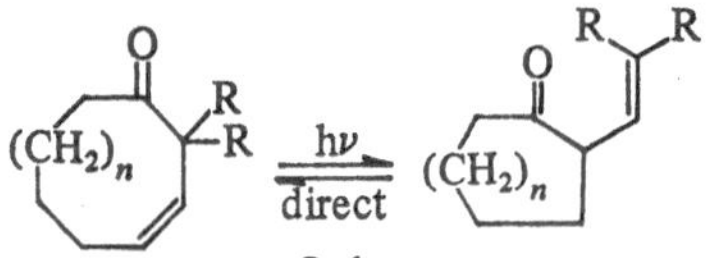

18

(11) [25,26]

19 n = 0, 1
    R = H, CH₃

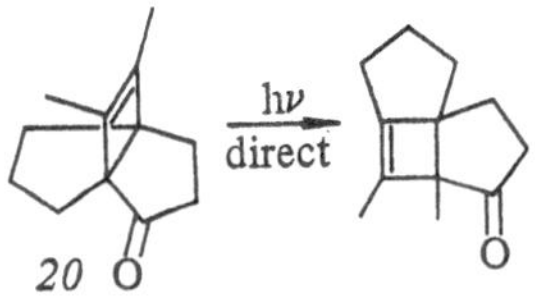

(12) [27]

20

Various factors can facilitate the 1,3-shift (*i.e.* will increase the quantum yield of the process). One factor can be the presence of a substituent on the α-carbon (atom 2 in *1*), which can stabilize an incipient allyl radical (see Section I.4 for the mechanism of the reaction). Geometric arrangement of the C=C and C=O bonds such that there is increased interaction of the two moieties (reflected by enhancement of the n-π* absorption) [3] will also facilitate the rearrangement. Susbtitution on the carbon-carbon double bond can also help the rearrangement by increasing orbital mixing and stabilizing the incipient allyl radical (for example, by methyl which indeed gives extra n-π* enhancement.) [19b] It is possible to rationalize the composition of a number of photoequilibria with the aid of the factors mentioned.

In a few cases 1,3-acyl shifts are found which apparently stem from the triplet state. The 1,3-acyl shift product *14* obtained in the sensitized

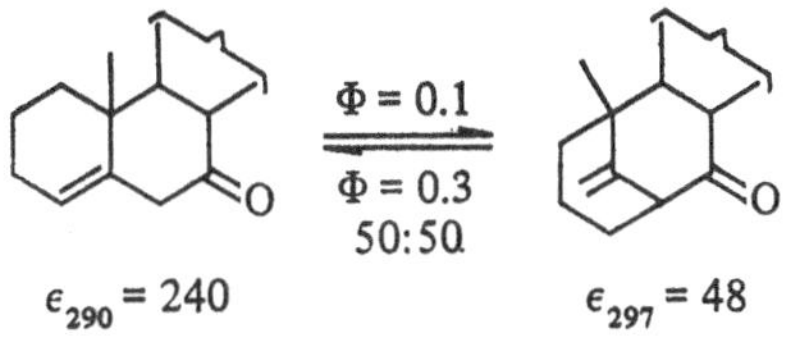

$$\Phi = 0.1 \quad \Phi = 0.3 \quad 50:50$$

$\epsilon_{290} = 240$ ⇌ $\epsilon_{297} = 48$ (13) [28]

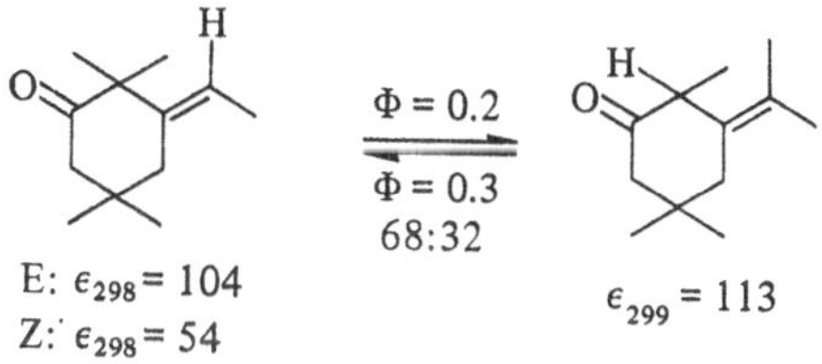

$\Phi = 0.2$ ⇌ $\Phi = 0.3$ 68:32 (14) [29]

E: $\epsilon_{298} = 104$
Z: $\epsilon_{298} = 54$ $\epsilon_{299} = 113$

irradiation of *13* was proven to be indeed a triplet product. [18] In the special case of aldehyde *21* where the carbon-carbon double bond is part of an α,β- enone the rearrangement product *22* was shown to stem from the triplet as well as the singlet state of *21*. [30] The chromophore in *24* being related to that in *21* suggests that *26* might also arise *via* the triplet.

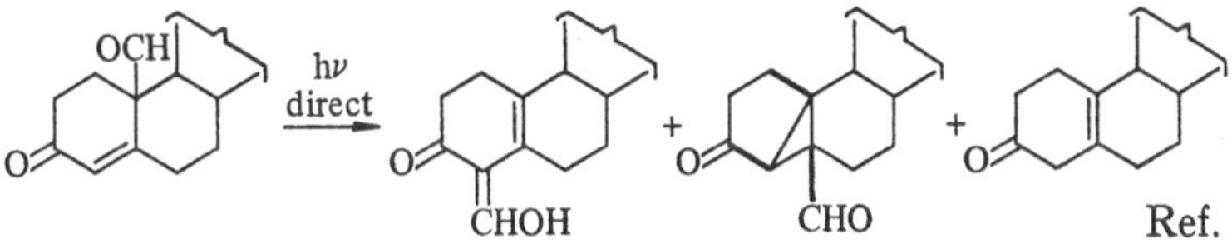

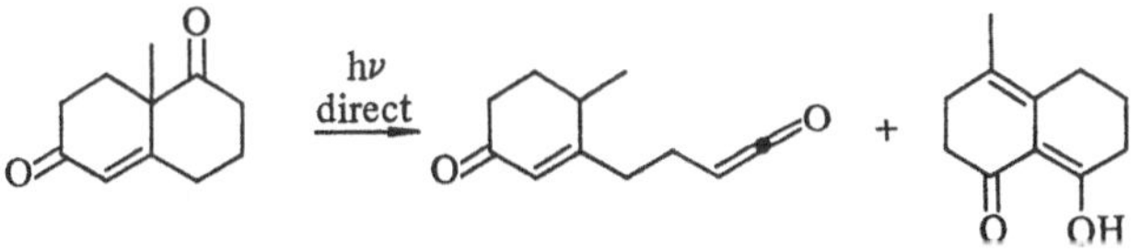

21   22   23   Ref. [30]

24   25   26   Ref. [30]

Triplet 1,3-acyl shifts are also found in the bicyclo[3.2.0]-hept-6-en-2-one series. Unexpectedly, compound *27* on sensitization with acetone does not give the 1,2-acyl product at all, but the 1,3-product instead [compare with Eq. (1)]. [31,32] Interestingly enough, *28* [33] and *29*, [31]

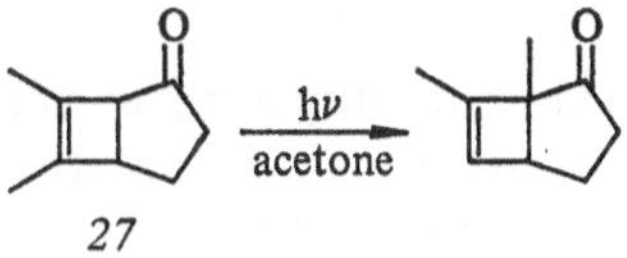

27   Ref. [31]

in which the carbonyl group of the β,γ-unsaturated ketone is part of an α,β-unsaturated system, on acetone sensitized irradiation, just like in the direct irradiation, give only 1,3-acyl shift products. In some sensitized irradiations it is not always clear whether the 1,3-acyl shift products,

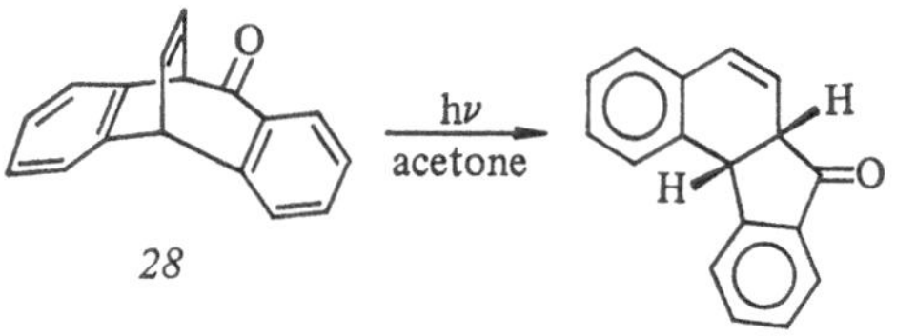

Ref. [33]

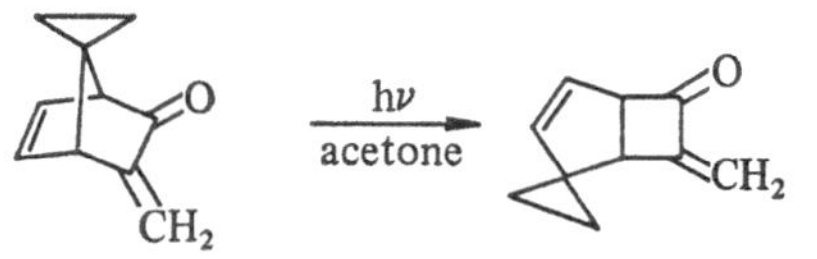

28

Ref. [31]

29

which are also formed on direct irradiation [*31* Eq. (15)], [34] really stem from the triplet state or from the singlet state populated by residual direct irradiation or maybe singlet sensitization. Quenching experiments are necessary to establish this point.

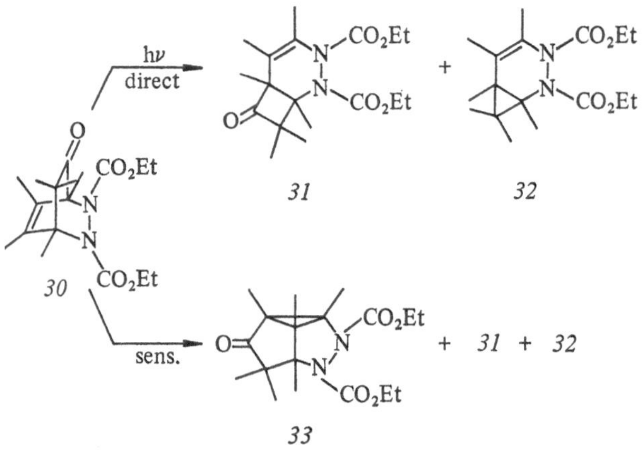

+ *31* + *32*   (15) [34]

2. Oxa-di-π-methane Rearrangement

The examples of 1,2-acyl shifts displayed at the lefthand side of the Eqs. (1—7) and in Eqs. (16—18) show that the oxa-di-π-methane rearrangement as a reaction of the triplet state of β,γ-unsaturated ketones is quite general. This specific rearrangement process has recently been reviewed. [35]

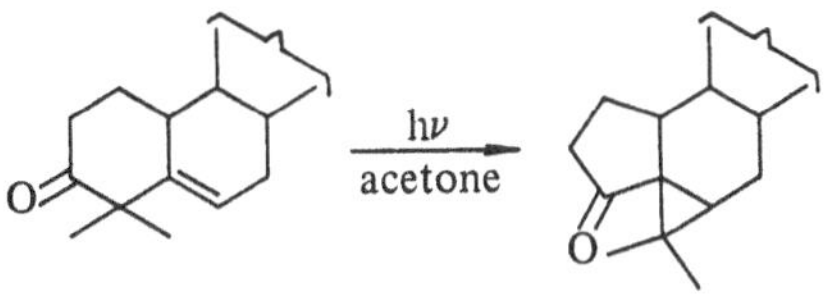

$$(16)\ [36]$$

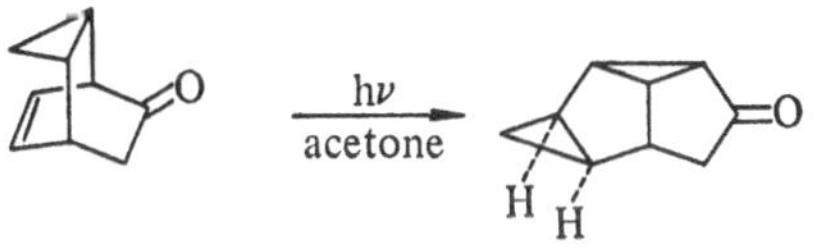

$$(17)\ [31]$$

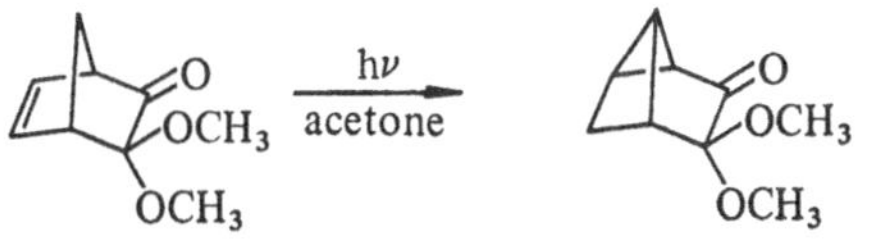

$$(18)\ [37]$$

34

Irradiation in benzene of *35* and *36* gives rise to double oxa-di-π-methane rearrangements. [38,39] Irradiation of *21* provides an example

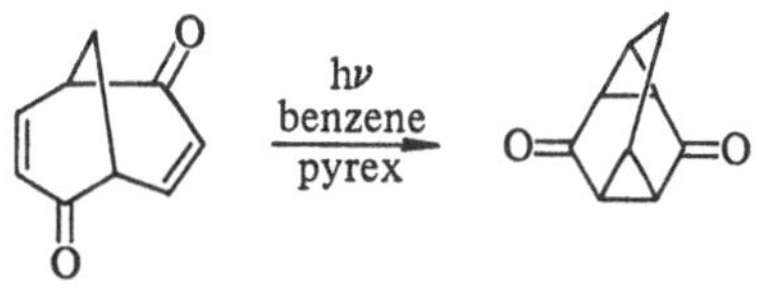

35

Ref. [38,39]

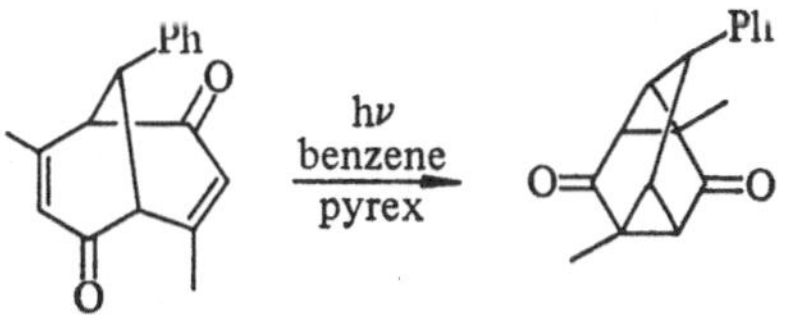

36

of a β,γ-unsaturated aldehyde which gives (in addition to other processes) a 1,2-acyl shift (to *23*). [30] Quenching studies demonstrated that all the reactions of this compound stem from the triplet state. The 1,2-acyl shift of *37* to *38* and *39* on direct irradiation is also shown to occur from the triplet state by quenching experiments. [40]

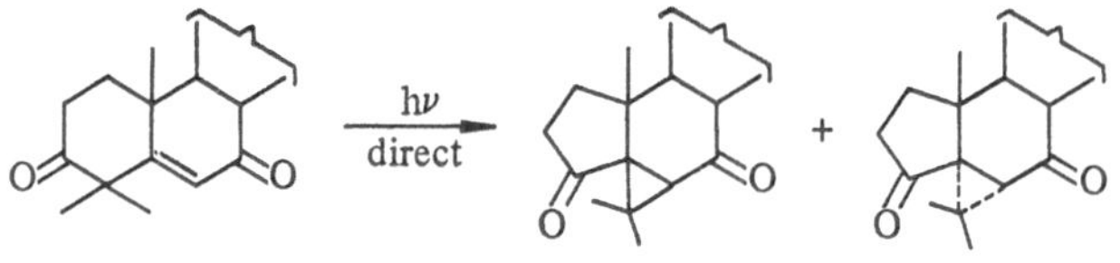

Ref. [40]

37   38   39

The photoisomerization of *40* and *41* shows that oxa-di-π-methane rearrangements are probably not restricted to β,γ-unsaturated ketones or aldehydes, in that other carbonyl containing chromophores can also undergo this type of reaction. [42,43]

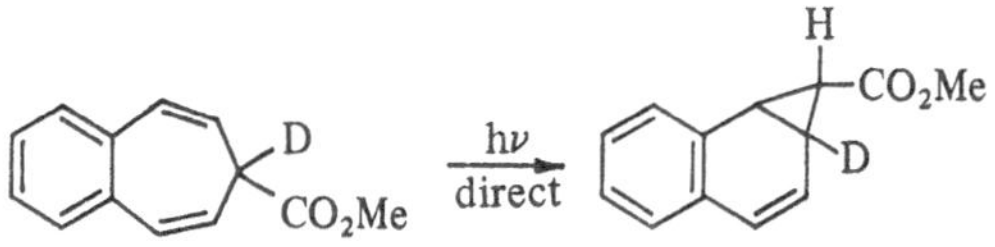

Ref. [42]

40

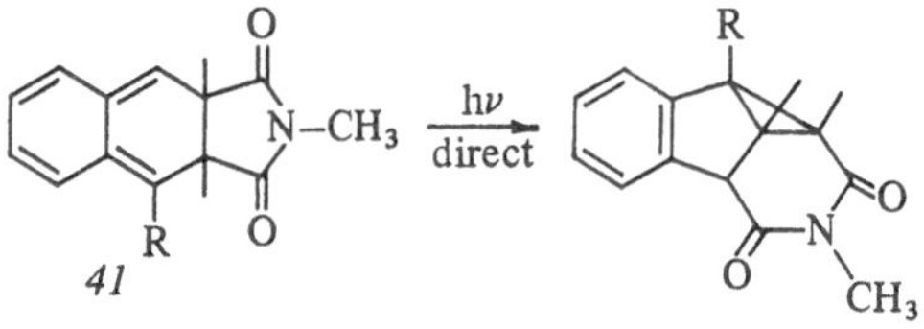

Ref. [43]

41

a) R = H   b) R = Ph

In a number of compounds there exists the possible choice between an oxa-di-π-methane and a di-π-methane rearrangement. Both modes of reaction are found. With *15c* [Eq. (7)] only the oxa-di-π-methane product *17c* is obtained and not the corresponding di-π-methane product. [20] On sensitized irradiation of *42* with R=H, *44a* is formed, which is the 1,2-acyl shift product and not *44b* and/or *44c* which are the benzo-vinyl-methane rearrangement products. [44] On the other hand, *45* gives only *46*, a di-π-methane reaction in which one of the participating double bonds is part of an α,β-enone, and none of the oxa-di-π-methane or benzo-vinyl methane products *47* and *48*, respectively, are formed.[45]. In the same way *49* only gives *50*, and not the oxa-di-π-methane or the divinylmethane product. [45] In the reaction of *51* to *52* it has not yet been established whether an oxa-di-π- or a di-π-methane pathway is taken. [46]

In the literature a number of cases have been reported where the oxa-di-π-methane rearrangement occurs on direct irradiation and where no studies are available to indicate that *isc* preceeded the rearrangement. [25,43,47,49,50,55,93] Among these is the first example of an oxa-di-π-methane rearrangement discovered in 1966. [47] A labeling experiment

Ref. [44]

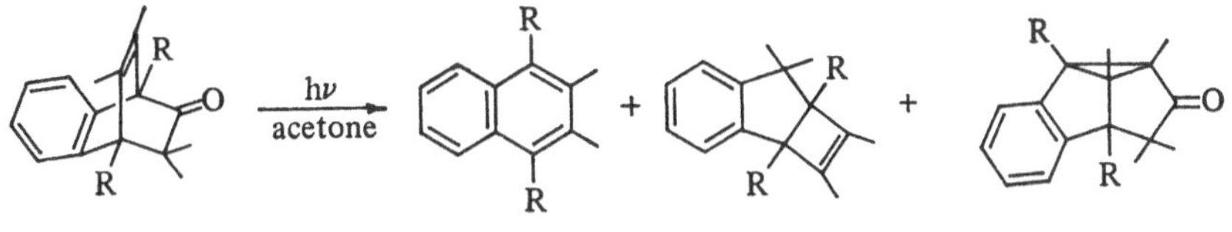

R = CH₃, H
42

43

44a

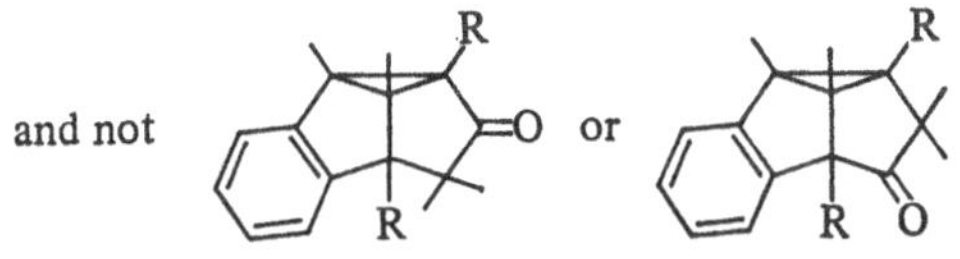

and not

44b   44c

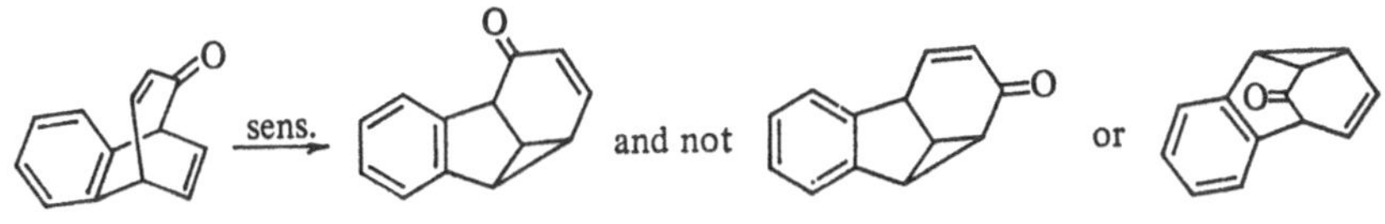

45   46   47   48

Ref. [45]

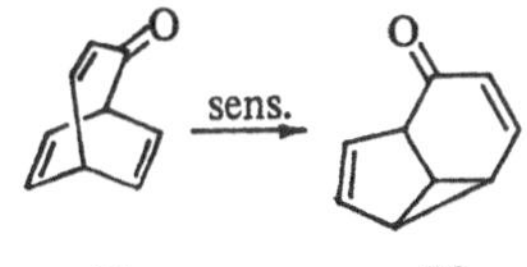

49   50

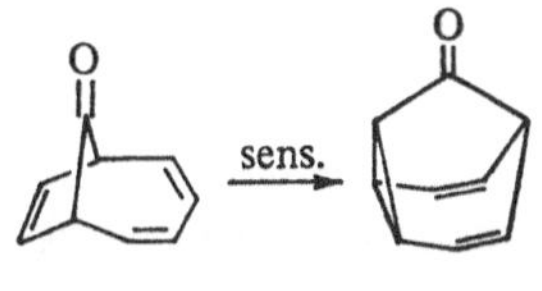

Ref. [45]

51   52

Ref. [46]

with ^{14}C indicates that the rearrangement of 53 to 54 indeed goes via migration of the benzoyl group. By analogy with 138, a compound of comparable structure where cis-trans-isomerization on direct irradiation has been shown to proceed via the triplet state, [48] it is plausible that the acetophenonic part of the chromophore provides a mode for *isc* to occur.

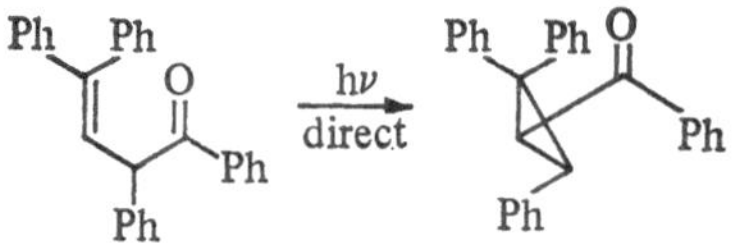

A most interesting case of the oxa-di-π-methane rearrangement, proceeding on direct irradiation, is found in *55* [49,50] and its steroid analogue *56*. [50] The 1,2-acyl shift occurs on direct irradiation, cannot

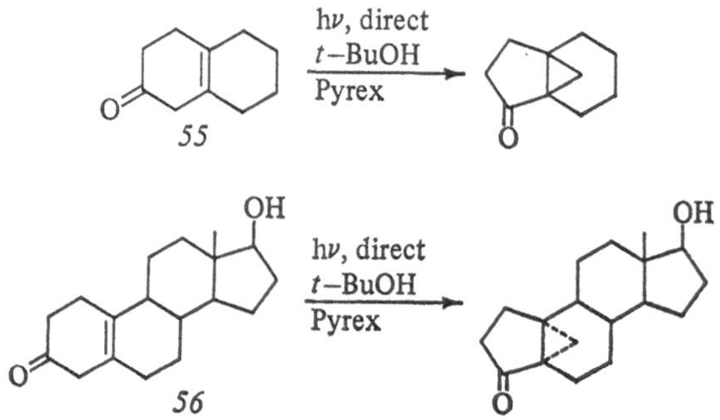

be sensitized by benzophenone, [50] but can be sensitized by acetone. [51] Quenching studies show that ϕ_{isc} of *55* ≥ 0.5. [51,52] This is very intriguing, because the isomeric steroid *57* does not show rearrangement [50] and the related compounds *58a—c* [51] and *59* [53,54] show the normal 1,3-shift on direct irradiation. The chromophore in *55* shows only a small enhancement of the n—π* absorption ($\varepsilon_{281} = 38$), [49] indicating that the dihedral angle between the carbonyl group and the ethylenic bond will be near 180°, in contrast with *58* ($\varepsilon_{292} = 252$) where the dihedral angle must be far from 180°. [52] (See earlier discussion). It has been suggested that therefore in *58* the orbital formed by α-cleavage (leading to 1,3-acyl shift) will overlap much more with the olefinic double bond than in *55*. In *55*, 1,3-shift will therefore be much slower than in *58* and *isc* gets its chance. [52] This explanation can be only part of the truth because other compounds with the same low n-π* extinction coefficient like *19* ($\varepsilon_{282} = 41$) [25,26] show the normal 1,3-shift behavior.

A further group of compounds which show the 1,2-acyl shifts on direct irradiation are those compounds which contain a O=C—C=C—C—C=O chromophore where the alkene part of the β,γ-unsaturated ketone is

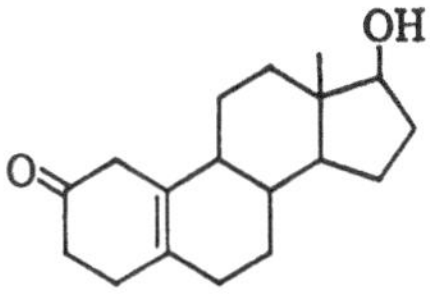

| 57 | 58a, R_1 = CH_3, R_2 = H | 58c | 59 |
| | 58b, R_1, R_2 = CH_3 | | |

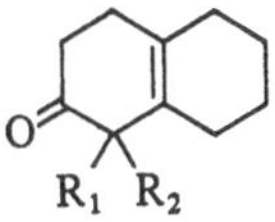

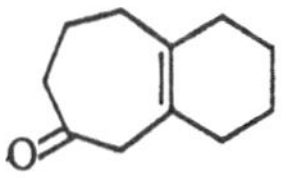

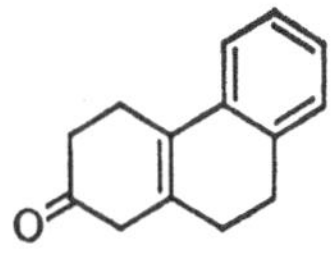

also part of an α,β-enone. [31,40] In the cases [31] where the multiplicity was studied, the 1,2-acyl shift was proven to stem from the triplet, the extended chromophore showing efficient *isc*. It is probably safe to assume that this behavior of the O=C—C=C—C—C=O grouping is general.

Relatively few examples exist in which the 1,2-acyl shifts possibly occur from the excited singlet state and not from the triplet state. [25, 55,57] In all cases the chromophore is not a simple undistorted β,γ-unsaturated ketone. The reaction of *60* to *61* can be sensitized but not

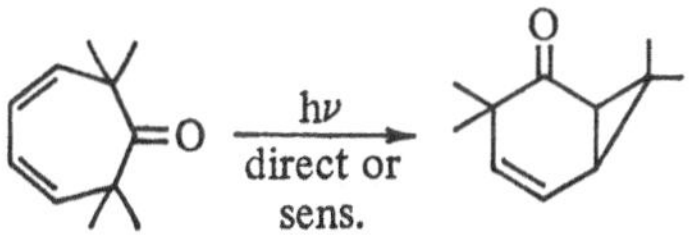

Ref. [25]

be quenched. [25] The 1,2-acyl shift on direct irradiation must therefore start from S_1 or from a short living unquenchable triplet. The same phenomena are found in the related compound *62*. [55,56] Both direct and sensitized irradiation in MeOH or CH_3CN give *63*. Various triplet quenchers do not retard the rate of formation of *63*. Irradiation in cyclohexane gives rise to the dimer *64*. On the basis of these results it has been suggested that in polar solvents the intersystem crossing is more efficient than the loss of carbon monoxide by a singlet process and it is the n,π*-triplet which leads to the rearrangement. In nonpolar solvent, it has been suggested that a twisted π,π*-triplet is of lower energy and it attacks the ground state molecule to give a dimer. No justification for the triplet assignments is given and thus should be regarded as an arbitrary assignment.

Strong indications for a singlet 1,2-shift are found in *65*. Cyclohexadienone *65* gives on irradiation in trifluorethanol the product *66*; in nonnucleophilic solvents *65* is inert. Unlike in the cases discussed later (Section III) the transformation of *65* to *66* in CF_3CH_2OH does *not*

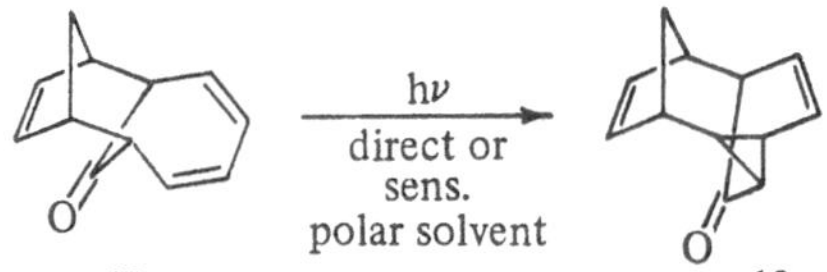

62   63

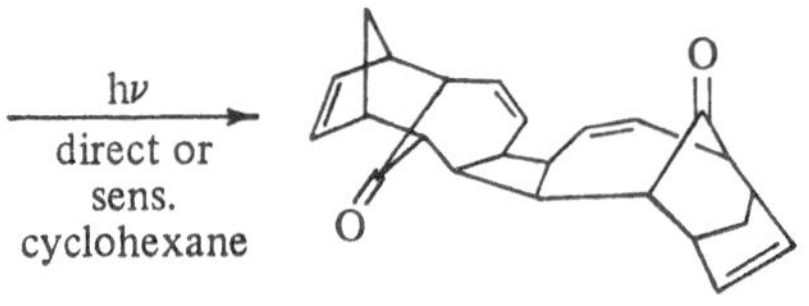

64   Ref. [55]

involve a ketene intermediate. Quenching and sensitizing studies show
that the reaction proceeds from the singlet state. As the influence of
the solvent on the UV spectrum of 65 shows, apparently the lowest
excited singlet in CF_3CH_2OH is the $\pi-\pi^*$ type, which gives the formal
1,2-shift product and not the $n-\pi^*$ type, which would give rise to ketene
formation. [57]

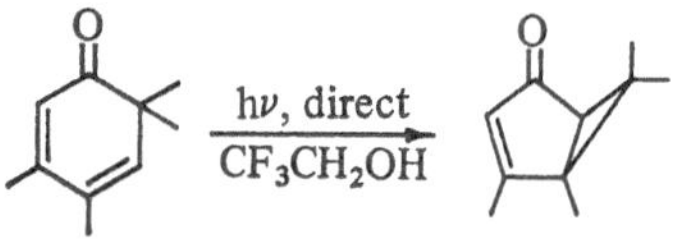

65   66   Ref. [57]

3. 1,3-Versus 1,2-Acyl Shift in Singlet Versus Triplet Behavior

From the above discussions it is apparent that the original generalities
about the photorearrangements are only a crude first order approximation. However, at least excited singlet states often give 1,3-acyl shifts
and triplets give 1,2-acyl shifts and only rarely are 1,3-acyl shifts from
the triplet and 1,2-acyl shifts from the excited singlet found. Two
different concepts have been proposed to account for the specificity of
the processes.

CNDO/S calculations with configuration interaction [7] for trans-
hex-4-en-2-one, in which the chromophore is geometrically arranged
as in bicyclo[2.2.2]octenone, give a lowest excited singlet state (of
$n-\pi^*$ character) in which in the β,γ-unsaturated chromophore system
an electron has been removed from an orbital which is strongly C—1,C—2
bonding and placed in one which is C—1,C—2 non-bonding and weakly,

C—1,C—4 bonding but C—1,C—3 anti-bonding. This indicates that α-cleavage and/or C—1,C—4 bonding leading to a 1,3-acyl shift will be favored from the excited singlet state.

The n—π* triplet has configurational properties similar to the n—π* singlet, and thus should be similar in its reactions. The lowest triplet of this system though is calculated to be of π—π* character. In this triplet, an electron is placed in an orbital which is C—1,C—3 bonding, but C—1, C—4 anti-bonding, with little weakening of the C—1,C—2 bond. So 1,2-acyl shift is indicated from the π—π* triplet state. In practice the lowest triplet state of β,γ-unsaturated ketones is indeed found to be of π—π* character. [5,6)]

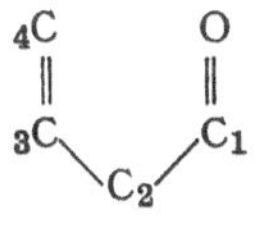

In the π—π* triplet the carbon-carbon double bond is calculated to be much more weakened than in the singlet excited state. *Cis, trans*-isomerization is therefore expected to be more efficient in the triplet state than in the singlet excited state. This is indeed found to be the case. [9,18,30,48,94)]

A different rationalization based on spin density distribution was proposed. [58)] In an n,π* excited state the spin polarization of the σ framework will result primarily from interaction with the singly occupied p_y orbital on oxygen (the singly occupied π* orbital is orthogonal to the framework). This will result in a spin distribution as in *67* (p_y orbital arbitrarily assigned α spin). Product formation is assumed to formally

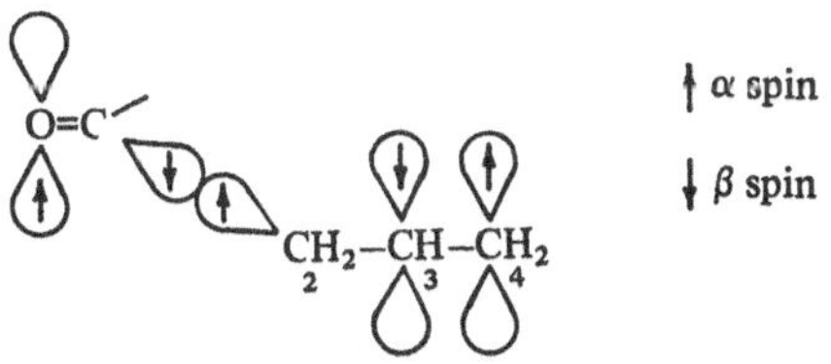

67

occur by electron pairing between the π* electron on the carbonyl carbon and electron spin density of opposite sign on one carbon of the olefin moiety. If the carbonyl electron is of α spin (*i.e.* species *67* is triplet state) migration of the acyl group to C—3 (*i.e.* 1,2-acyl shift) is anticipated; if the electron is of β spin (*i.e.* species *67* is singlet) migration to C—4 (*i.e.* 1,3-acyl shift) will occur. This second explanation is based on n—π* excited states; it is difficult to evaluate its value for the triplet, where reactions probably start from a state of π—π* character.[5,6]

4. Mechanism of the 1,3-Acyl Shift

Various mechanisms for the 1,3-acyl shift have been proposed, the following three pathways being the most widely discussed: a) a Norrish type I cleavage to an acyl and an allyl radical followed by recombination either to starting ketone or to the rearranged ketone [Eq. (19)]. [19b,21,25,33, 37,59,95,96] For some cases it has been pointed out that the radical centers do not become independent of each other but exist as a radical pàir in a solvent cage. [21,37,59] b) a concerted process as depicted in Eq. (20); [60] and c) a concerted $\pi2 + \sigma2$ cycloaddition of the C—1,C—2 bond with the C—3,C—4 double bond; [Eq. (21)]. [60] Mechanisms (b) and (c) have not always been properly distinguished. Circumstantial evidence has been reported which supports the nonconcerted and/or

a)

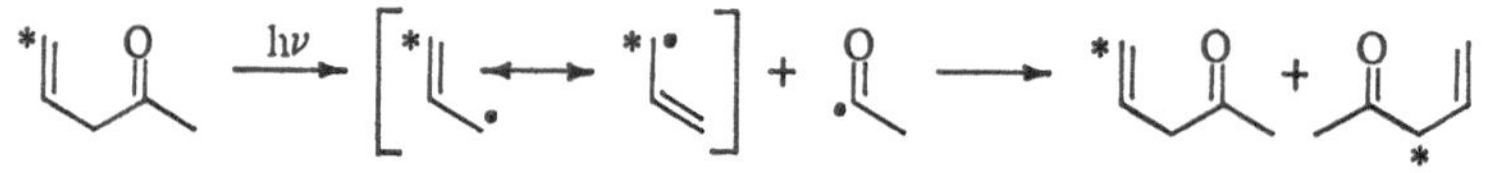

$$(19)$$

b)

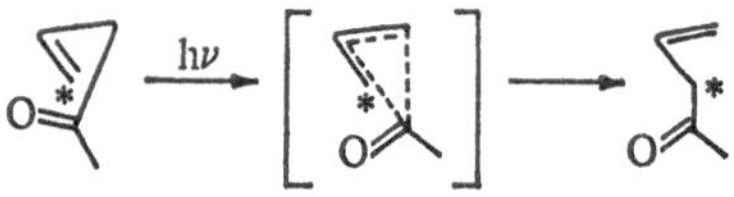

$$(20)$$

c)

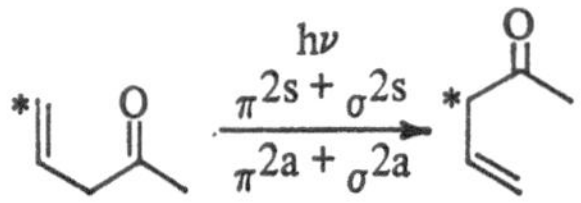

$$(21)$$

the concerted pathway. [15,16,17,28,53] The results from a study of compounds *68*, [28] *71* [53] and *74* [53] are the most relevant. Compound *68* gives on direct irradiation compound *70*. In this reaction, which is reversible, the identities of the 6α and 6β hydrogen atoms are preserved. [28]

The same phenomenon is found in *71* and *74*. [53] On irradiation of both compounds, 1,3-acyl shift products are formed in which the 1α-methyl and 1β-methyl group have not lost their identity, (*71a→72a*, etc.). Of the aldehyde products, which are formed concurrently, *73* is

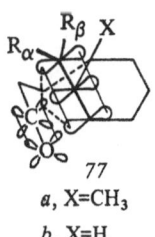

68   69   70

71   72   73

a, R_α=CD$_3$, R_β=CH$_3$ a, R_α=CD$_3$, R_β=CH$_3$ a, R =CD$_3$, Y=H

b, R_α=CH$_3$, R_β=CD$_3$ b, R_α=CH$_3$, R_β=CD$_3$ b, R =CH$_3$, Y=D

74   75   76

a, R_α=CD$_3$, R_β=CH$_3$ a, R_α=CD$_3$, R_β=CH$_3$ a, R=CD$_3$, Y=H

b, R_α=CH$_3$, R_β=CD$_3$ b, R_α=CH$_3$, R_β= CD$_3$ b, R=CH$_3$, Y=D

77

a, X=CH$_3$

b, X=H

also formed stereospecifically, but 76 is formed non-stereospecifically, 71a→73a but 74a→76a+76b.

These results have been rationalized as follows. [28,53] The 1,3-acyl shifts proceed through a radical pair like 69 or 77. This radical pair must be very tight in order to preserve the observed stereospecificity. The tightness can be provided by solvation and residual interaction of the parts of the radical pair. A portion of the solvated intimate radical pair is liberated from the solvent cage (and residual interactions) to form free biradicals which give aldehydes *73* and *76*. The allyl radical part of *77a* is hindered in rotation by the angular methyl group and thus gives aldehydes *73* via specific abstraction of β-methyl hydrogens by the acyl radical; *77b* is free to rotate and gives a 1:1 mixture of *76a* and *76b*. Scrambling of the CH$_3$ and CD$_3$ groups in recovered *74* has not occurred so the free radical intermediate reverts neither to *74* nor *75* but gives the aldehyde instead.

An alternative explanation is that the 1,3-acyl shift occurs through a concerted process, which also accounts for the stereospecificity. If such be the case, a competitive reaction of the excited singlet state is α-cleavage to the intimate and/or free radical pair, which goes on to the aldehyde. The second explanation looks slightly more attractive; it is not too easy to understand why a free biradical would give only aldehyde formation and no recombination to starting material and 1,3-acyl shift product. When in the intimate singlet biradical, proposed to go on directly to 1,3-shift product, there would still be appreciable interaction between the two radical centers, a discussion about differentiation between a biradical or a concerted mechanism becomes senseless.

An interesting aspect of the above hypothesis is the involvement of an oriented radical pair as an intermediate whose limiting form could be, in fact, the same as an concerted process, seems to be borne out by CIDNP experiments. [61] Whereas in *12* the radical pathway is at most a minor one, in *78* the radical pathway is the major and possibly only one.[69] The nature of the stabilizing group defines the place of the radical pair on the continuum of free radical pair to concerted process.

12   *78*

5. Mechanism of the 1,2-Acyl Shift

Various mechanisms for the 1,2-acyl shift have been proposed, the following three pathways being the most widely discussed: a) a Norrish type I cleavage to an acyl and an allyl radical followed by recombination of the radical pair to a cyclopropyl ketone or closure of the allyl radical to a cyclopropyl radical, which then recombines with the acyl radical [Eq.

(22)]; [36,47,49,50] b) The oxa-di-π-methane type, *i.e.*, initial bonding between the carbonyl carbon and the β-carbon, followed by rearrangement or a concerted version of these steps ($_\pi 2 + _\sigma 2 + _\pi 2$) [Eq. (23)]; [9,11, 14,16,20,31,40,41,74] and c) a concerted $_\pi 2 + _\sigma 2$ cycloaddition of the carbonyl-α-carbon bond with the olefinic bond [Eq. (24)]. [17,40,61,65,74]

a)  $$\qquad\qquad\qquad\qquad\qquad\qquad\qquad\qquad (22)$$

b)  $$\qquad\qquad\qquad\qquad\qquad\qquad\qquad\qquad (23)$$

c) 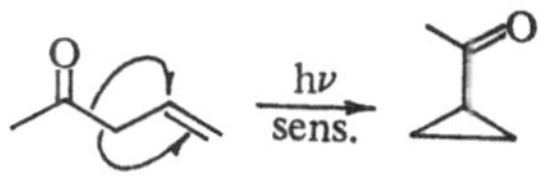 $$\qquad\qquad\qquad\qquad\qquad\qquad\qquad\qquad (24)$$

Mechanism (*a*) does not seem very likely. Of the various pathways available to the radical pair, bonding to the central atom of an allylic system or rearrangement of allyl radical to cyclopropyl radical are energetically the least favorable. [62] Closure to either starting material or isomeric ketone would be expected to take precedence. Neither isomeric β,γ-unsaturated ketone nor scrambling of the substituents on the methane carbon atom in residual starting material is found.

Mechanism (*b*), the oxa-di-π-methane mechanism, is in line with CNDO calculations which show that π-orbital interaction between the carbonyl and the β-carbon is favored in the π–π* triplet state. [7] The concerted process, mechanism (*c*) has been pointed out to be unlikely in that a cyclopropyl ketone triplet is of higher energy than the triplet of the starting ketone. [63] It is reasoned that the rearrangement thus cannot take place entirely in the triplet manifold and a spin inversion to yield an intermediate during the overall reaction is needed. [54] On the other hand, this explanation may be too simple, there is no *a priori* reason why a reaction of triplet state to ground state should not proceed with synchronous bond breaking and formation. [64]

Both the suprafacial, suprafacial (s + s) and antarafacial, antarafacial (a + a) pathways of the $_\pi 2 + _\sigma 2$ cycloaddition are photochemically allowed. [60,65] A (a + a) pathway should give inversion about the methane carbon atom, the (s + s) pathway retention. The 1,2-acyl rearrangement has been labeled $\pi 2_a + \sigma 2_a$ in a number of cases in the literature, [16,17, 20,55,66,67] however, in some cases the starting materials were unable to adopt the (s + s) pathway. For example, in an cyclohex-3-en-1-one the (s + s) mode would give a *trans*-junction between the five- and the three-membered ring of the product. In an acyclic β,γ-unsaturated ketone both the (s + s) and (a + a) modes are geometrically permitted.

The stereochemical results obtained in the various investigations pertinent to the establishment of the mechanism are equivocal. In *80*, obtained from the sensitized irradiation of *79*, the CH_3 and CD_3 groups are completely scrambled. [54] In the same fashion, the sensitized irradiation of optically active *81* gives the diastereoisomeric *82*, in which optical activity apparently is lost. [68] A partial stereospecificity is observed in *83* which gives on sensitization a 7:1 mixture of the *endo/exo* products *84* and *85*. [67,69]

All these results are in accord with a non-concerted mechanism. An alternative route by which the stereochemical identity of the methane

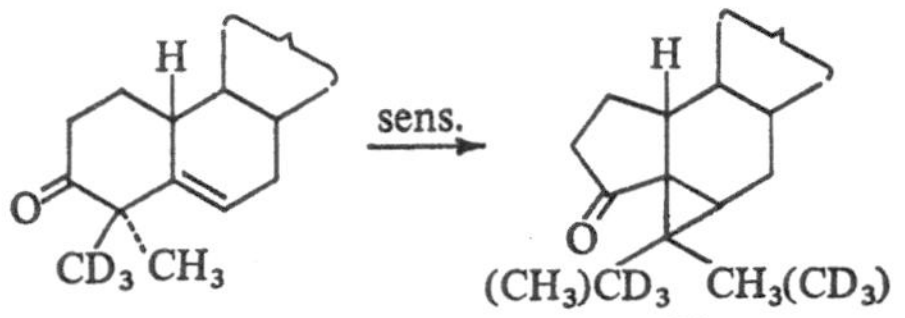

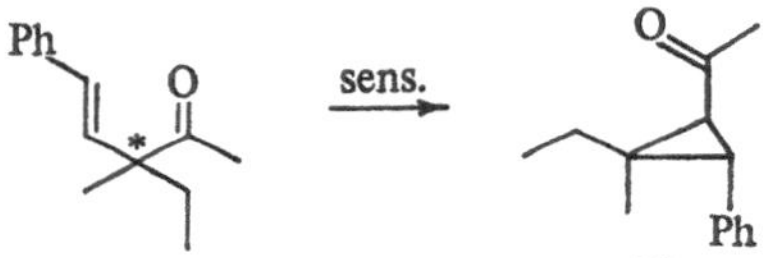

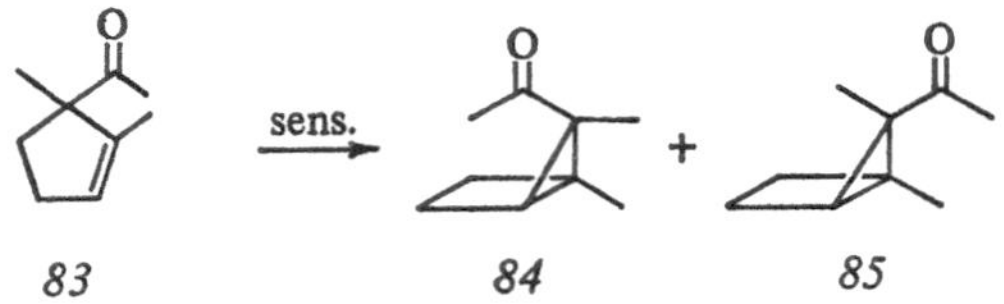

carbon atom could have gotten lost is by triplet isomerization of either the starting material or product. *Cis-trans* isomerizations of cyclopropyl ketones are known. [70] In the cases where this was investigated, neither racemization of the starting material nor isomerization of the cyclopropyl ketone product was found [68,69,71]

Total stereospecificity has been found in the unrestrained *86*; *86a* gives only *87a* and *86b* only *87b*, i..e, retention of configuration. [71a] It has been shown that isomerization does not occur in starting material

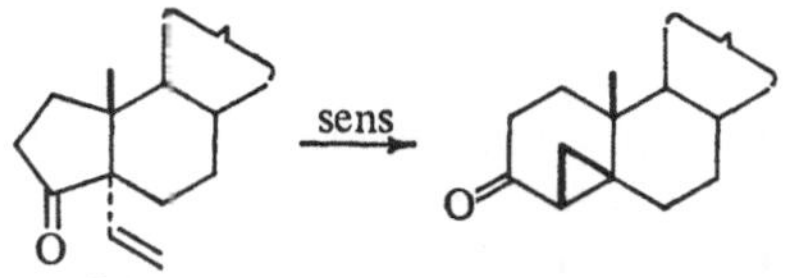

Ref. [71a]

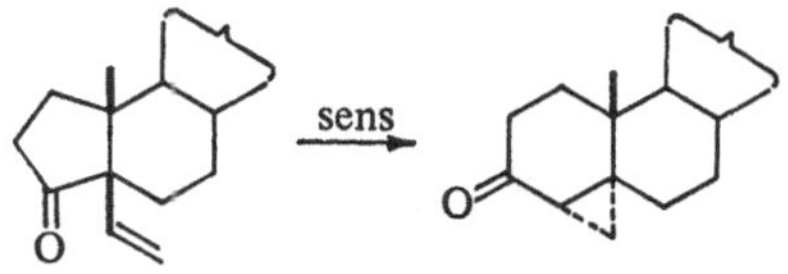

or product. [71a] These results are in accord with an allowed concerted $\pi 2_s + \pi 2_s$ process, or with a stepwise mechanism in which initial carbonyl-vinyl bonding only occurs in the preferred conformation of *86* in which the vinyl group points away from the steroid nucleus. This same conformation preference would inhibit a concerted $\pi 2_a + \pi 2_a$ process.

The foregoing study only evaluated the stereochemical fate of the σ-portion of the concerted process. The utilization of *86c* permits the evaluation of the stereochemical fate of both bonds and in contrast to

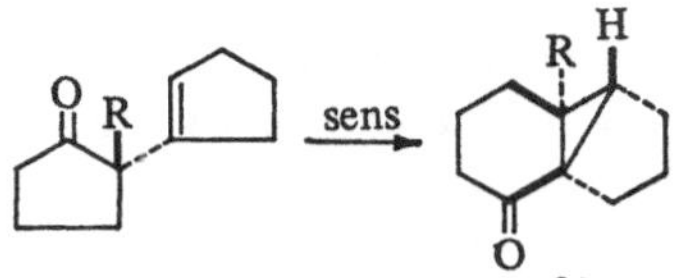

Ref. [71b]

R = CH₂COOH

the results found for *87a* and *87b*, the rearrangement product *87c* possesses a *cis* arrangement of the cyclopropane ring attachments and an inverted C—3 center. [71b] This result is in accord with an allowed concerted $\pi 2_a + \pi 2_a$ process. The finding of *complete* inversion casts great doubt on a stepwise process unless some specific conformational effects in the excited state has a large effect on the mode of vibrational relaxation of the triplet.

Discrepancies are also found in compounds which contain a disturbed β,γ-unsaturated ketone chromophore. Irradiation of *88* leads to the cyclopropyl ketones *89* and *90* in which the methyl groups, originally attached to the methane carbon, are completely scrambled. [40,41] This result is not in accord with a $\pi 2_a + \sigma 2_a$ mechanism. On the other hand, *91*, which contains the same extended β,γ-unsaturated chromophore as *88* gives *92* with the acyl group *endo*. [72a] This is in accord with a $\pi 2_a +$

93

$_{\sigma}2_a$ mechanism, unless *endo 92* would by chance be the thermodynamically most stable isomer. In both of these cases, however, it was not determined whether the product stereochemistry came during the formation of the product or whether the primary cyclopropane derivative was photochemically isomerizable. [70a)]

Ref. [40,41]

88 *89* *90*

Ref. [72a)]

91 *92*

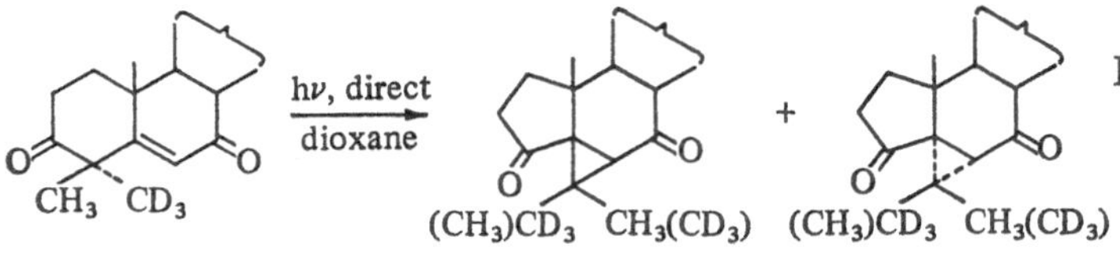

A different approach to the problem of concertedness has been pursued in the study of *7* vs *95*. [15,16)] Sensitized irradiation of *7* gives *9* and *10* [Eq. (2)]. To determine whether diradical *94* is on the reaction path of *7* to *10* as it should according to the non-concerted oxa-di-π-methane mechanism, *94* was independently generated by photolysis of *95*. All the components in the previous sensitized study [Eq. (2)], *i.e.* *7, 9, 10* were formed from the triplet *95*. However, *8* observed in the

7 *93* *94*

95 *8* 1%

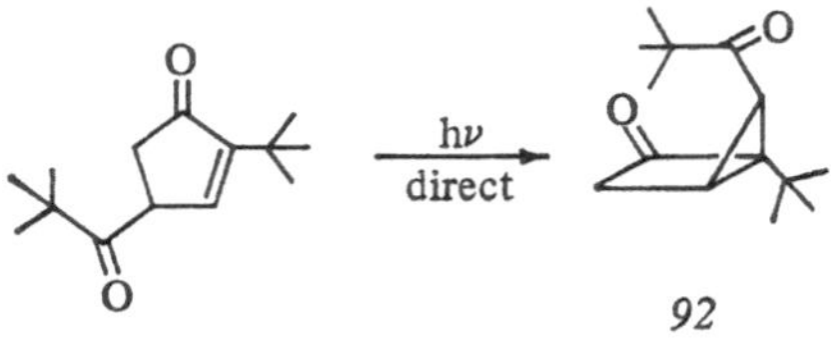

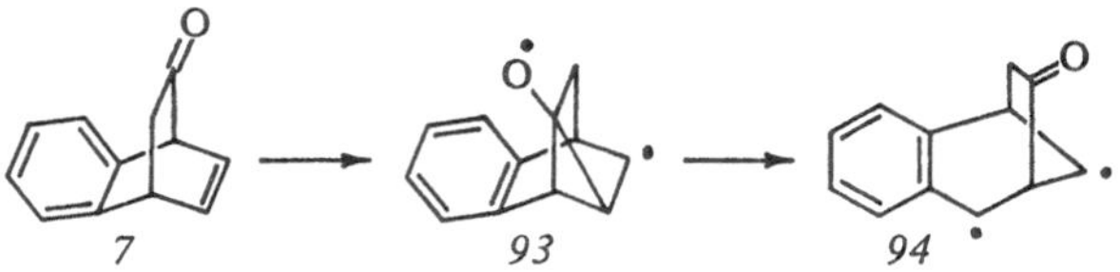

photolysis of the lactone *95* is not found in the sensitized irradiation of *7* Since *8* is not produced from *7*, it was concluded that the 1,2-rearrangement should, by default, be considered a concerted reaction. [16)]

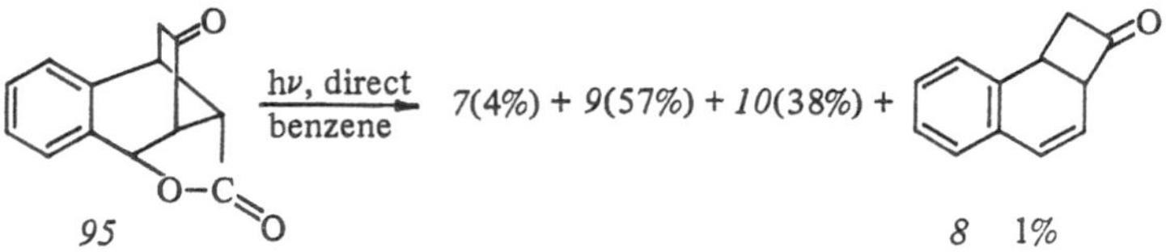

As is so often the case in photochemical reactions, single examples appear which tend to cast doubt on conclusions reached previously on the basis of a large volume of data. For the case at hand where an concerted process is indicated, the relationship of the reactions to the general 1,2-acyl rearrangement must be established.

II. Decarbonylation

The occurrence of decarbonylation is remarkable in the sense that β,γ-unsaturated ketones are the only class of ketones (with a few other special cases) which give decarbonylation on direct irradiation in solution. Decarbonylation sometimes accompanies 1,3-acyl shifts to a greater or lesser extent. [37,22,23,73,74,80] In a number of cases photodecarbonylation occurs to the exclusion of other processes. Equations (25) to (30) give some examples.

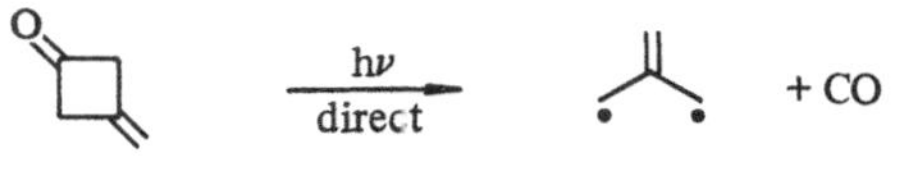

$$\text{(25) }[75]$$

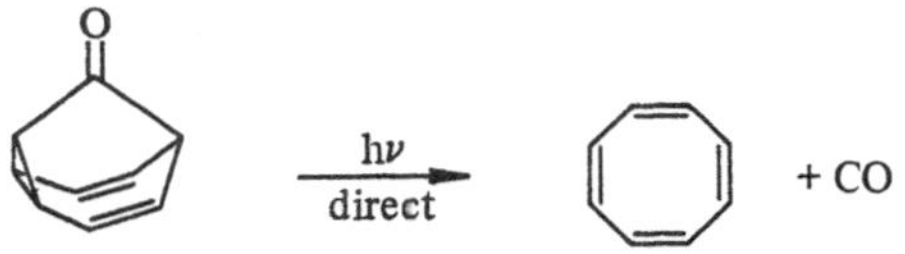

$$\text{(26) }[46,76]$$

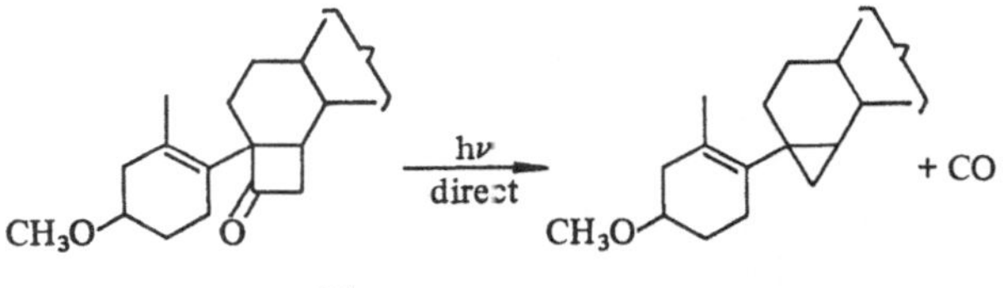

$$\text{(27) }[74]$$

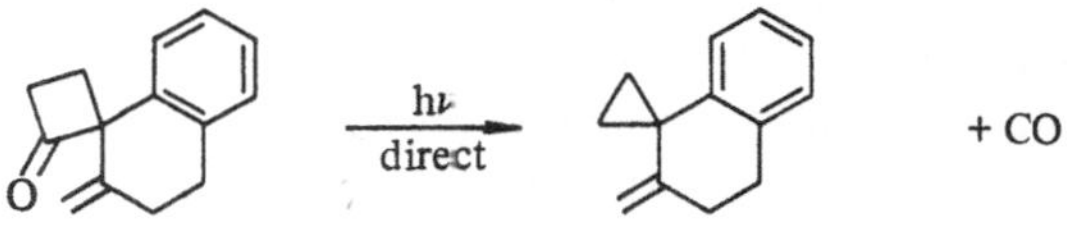

$$\text{(28) }[53]$$

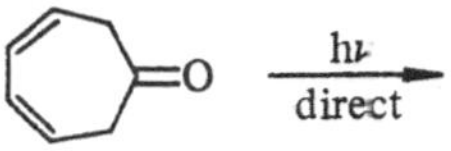

$$CH_2=CH-CH=CH-CH=CH_2 + CO \quad \text{(29) }[77-79]$$

96

97

98

95

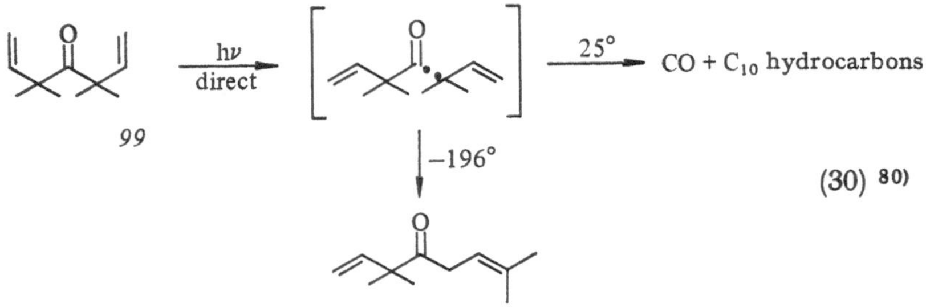

$$(30)^{80)}$$

Compounds with a 2-vinylcyclobutanone grouping like *16, 96, 97,* and *121* give either decarbonylation or ketene fragmentation. [19,20,53, 74,81] Substitution on the α-carbon atom seems to help decarbonylation: whereas *16* ans *121* give only ketene fragmentation, [19,20] *100* and *101* give only decarbonylation. [73,37]

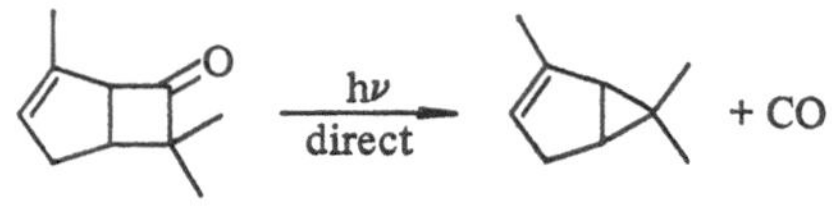

Ref. [73]

Ref. [37]

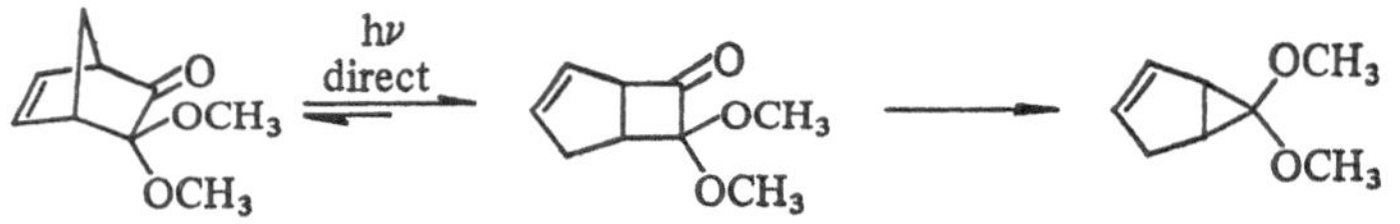

The behavior of *98*, which gives decarbonylation in a clean efficient reaction, is strikingly contrasted by that of the tetramethyl analogue *60*, which on direct irradiation gives no decarbonylation but only the 1,2-acyl shift product *61*. [25] (See Section I.2). As an explanation it has been proposed that the methyl groups in *60* inhibit the steric requirements for a chelotropic carbon monoxide expulsion. [25]

Singlet decarbonylation is also the prominent reaction in β,γ-unsaturated aldehydes in which the chromophore is not perturbed as it is in the O=C—C=C—C=O group. The efficiency of the reaction is found to be a function of the amount of orbital mixing of the aldehyde and ethylenic group. [82] Whereas *102* and *104*, which show enhanced n—π* absorption, give efficient decarbonylation reactions from the excited singlet state to *103, 105* and *106*, respectively, *107*, which has a non enhanced n—π* band, gives a sluggish reaction to a number of products, among which are *108* and *109* which stem from the triplet state. [82]

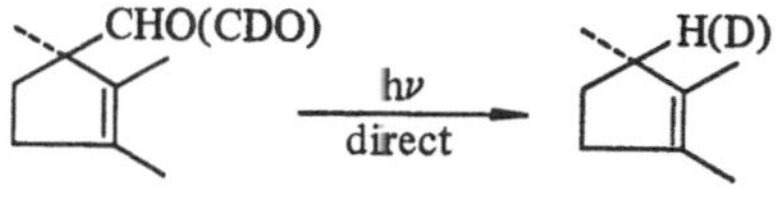

CHO(CDO)   H(D)

102 ϵ_{305}=135   *103*   Ref. [82]

CHO(CDO)   H(D)   CH$_3$(CH$_2$D)   Ref. [82]

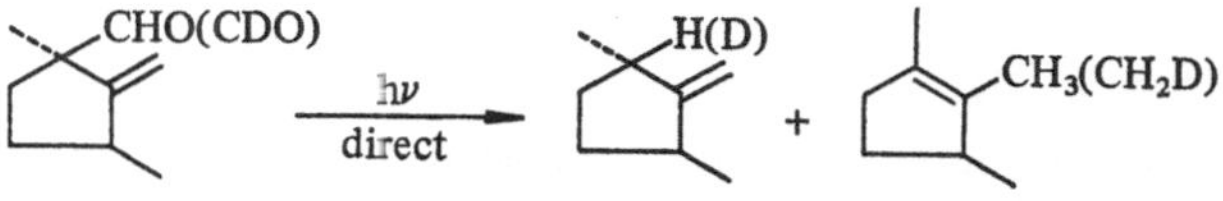

104 ϵ_{299}=68   *105*   +   *106*

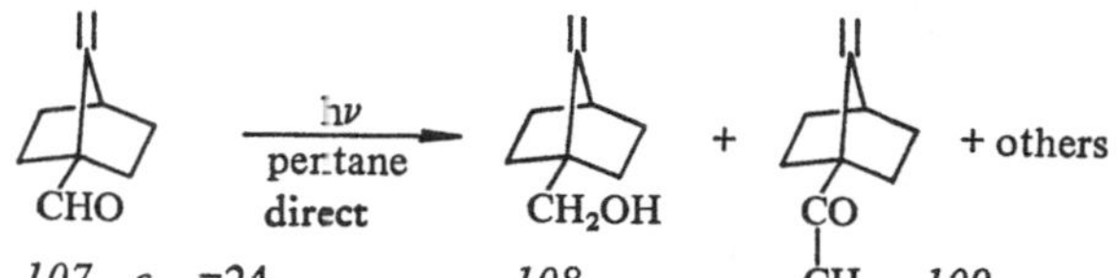

CHO   CH$_2$OH   CO   + others

107 ϵ_{302}=24   direct   *108*   CH$_3$ *109*   Ref. [82]

pertane

Cyclopent-3-en-1-ones are another subgroup of β,γ-unsaturated ketones in which decarbonylation is the almost exclusive process. These compounds show no n—π* enhanced absorption band.

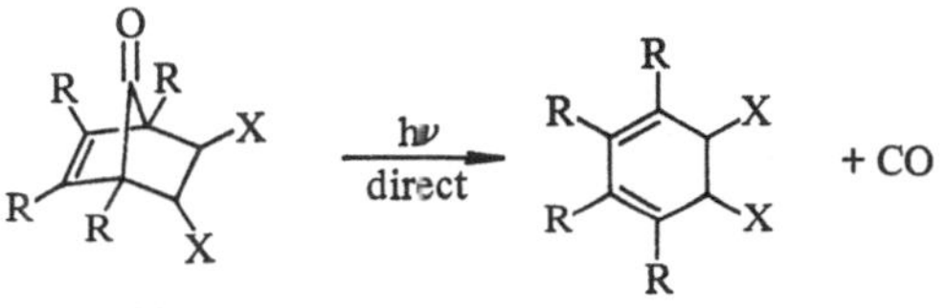

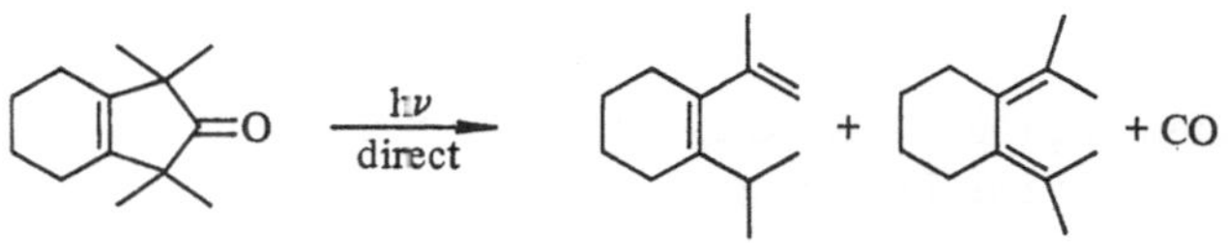

110   hν / direct   + CO   Ref. [83]

111   hν / direct   *112*   +   *113*   + CO   Ref. [84]

Compounds of the general tape *110*, with a variety of substituents all give an efficient singlet decarbonylation. [83] Ketone *111* gives an efficient carbon monoxide expulsion [84] (the hydrogen transfer to *112* is intramolecular) in contrast with *114*, which only decarbonylates sluggishly ($\phi = 0.02$) and mainly from the triplet. [85] Relief of strain, formation of stable molecules and stabilization of intermediates have been recognized as factors which form driving forces for the decarbonylation. [83,85]

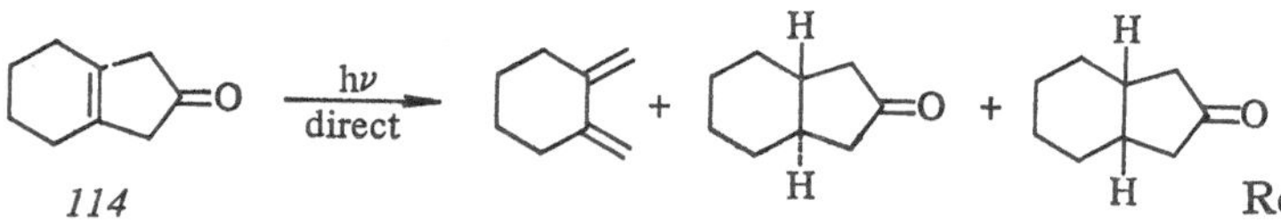

Both a one step and a two step reaction to a diradical intermediate or directly to product have been proposed as mechanisms for the decarbonylation. [60,83,84] The results for *115* and *116* show that at least in in this case the decarbonylation is a stepwise nonchelotropic reaction. [86]

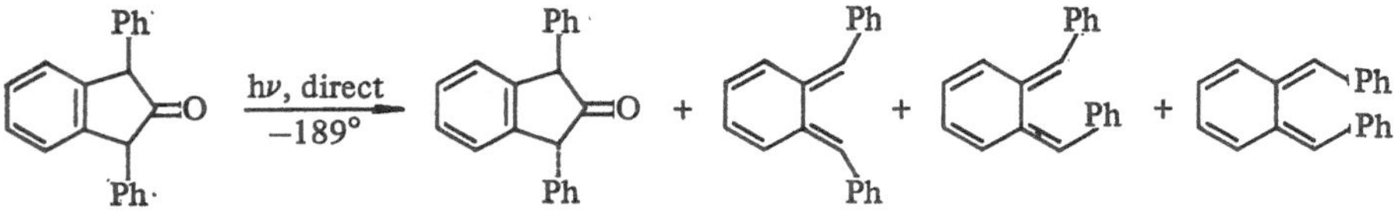

The different product ratios of the dienes formed in the decarbonylation of *cis-* and *trans 114a* resp. on the other hand are in line with a concerted pathway [86a].

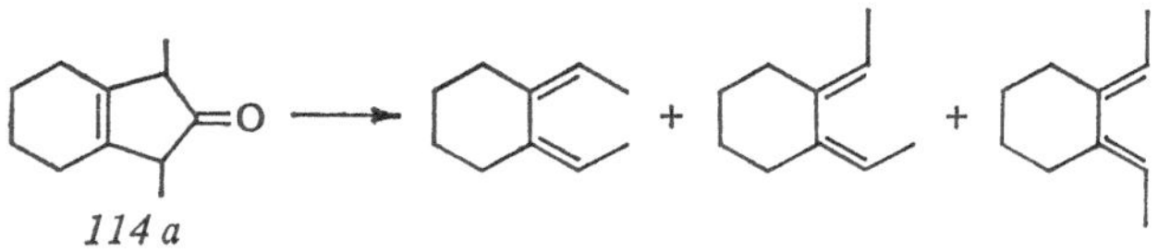

The finding of mixed decarbonylated products in the irradiation of *11* [9] and *99* [80] points out that in these acyclic compounds the decarbonylation also is not a chelotropic process.

Decarbonylation is as rare from the triplet state as it is abundant from the singlet excited state. It is found that triplet decarbonylation in *114* is an inefficient process and indeed compares to reduction. [85]

It is surprising that these two highly inefficient reactions are the only routes followed. It is to be noted, however, that triplet-induced 1,2-acyl shifts are only efficient when there is an enhanced n—π* band and such is not present in *114*. The steroidal β,γ-unsaturated aldehyde *21* undergoes decarbonylation from the triplet state less efficiently than the 1,2- and 1,3-acyl shift. [31] The related monocyclic aldehyde *120* only shows decarbonylation, and by analogy with *21* the reaction is suggested to stem from the triplet state. [31]

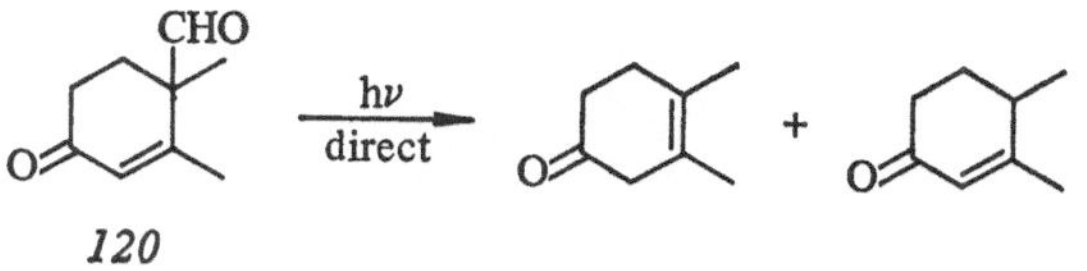

120   Ref. [31]

It is not clear whether the decarbonylation products *32* [35] and *43* [44] which are formed from *30* and *42* both in direct and sensitized irradiation, stem from the triplet state in the sensitized reaction or are the product of residual direct irradiation.

III. Ketene Formation

Ketene fragmentation is the prominent process in the photolysis of compounds with the general formula *121* $(n = 1,2)$. [15,16,19a,20,81]

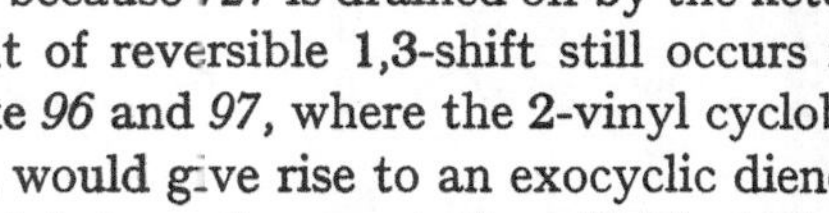

121

When these ketones are the product of a 1,3-acyl shift, a photoequilibrium is not established because *121* is drained off by the ketene fragmentation. A limited amount of reversible 1,3-shift still occurs in a few cases. [37] In compounds like *96* and *97*, where the 2-vinyl cyclobutanone group on ketene formation would give rise to an exocyclic diene, decarbonylation occurs instead of ketene fragmentation. [53,74] Although the detailed mechanism of the elimination process is not known, from the examples at hand it appears to be efficient only when extensive overlap with the developing orbitals occurs, *i.e.*, *121* is perfectly set up for overlap throughout the reaction, *96* and *97* are certainly not. Ketene formation is also prominent in the special ketones *122* [87] and *123*. [57,88] The

99

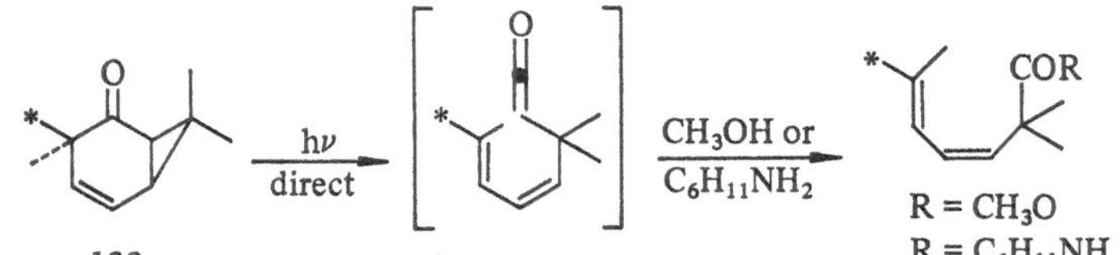

Ref. [87]

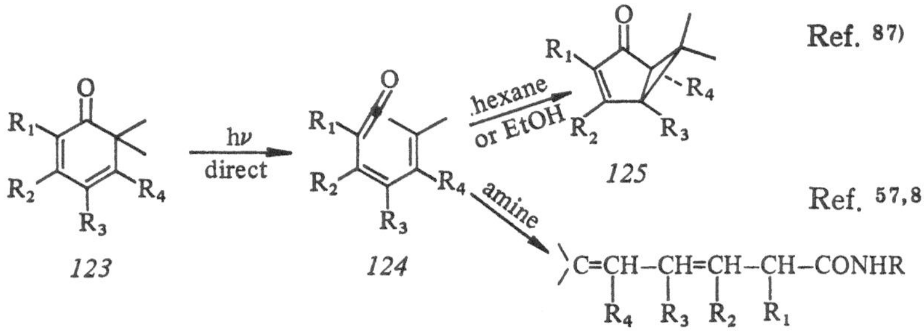

Ref. [57,88]

stereochemical identity of the methyl groups on the α-carbon atom of *122* is maintained in the reaction as indicated. [87] This fact points to a concerted process for the ketene formation. A very tight solvent caged intermediate diradical cannot, of course, be excluded.

The formal 1,2-acyl shift product *125* does not come directly from *123*. In the heavily methyl substituted cyclohexadienones *123* which have been intensively studied, the only photochemical reaction is the formation of ketenes *124*; [88] *125* is formed in a thermal reaction from *124*. Only nucleophiles stronger than methanol, like amines, are able to trap the intermediate ketene. [57,88]

No quenching and/or sensitization studies are available which establish with certainty ketene formation proceeding from the triplet state. The ketene formation from *24* and *126* most likely occurs from the triplet state, since in related systems with an α,β-enone as part of the chromophore, only reactions from the triplet are found. [31,90]

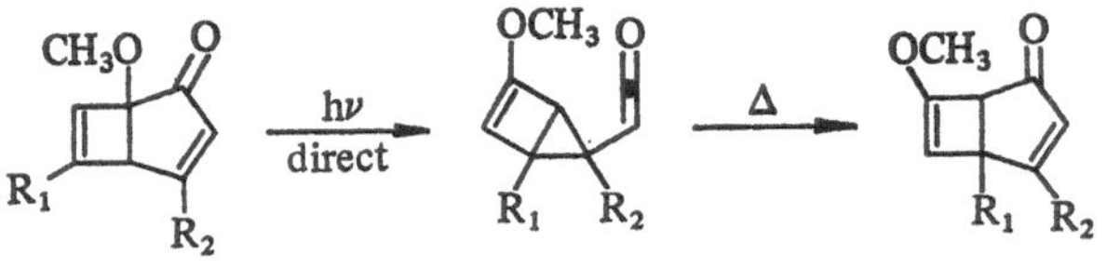

Ref. [90]

It is also not clear whether the naphthalenes formed in the sensitized irradiation of *7* and *42* are really triplet ketene fragmentation products or are the singlet products of residual direct irradiation. [13,15,16,44]

100

IV. Aldehyde Formation

As an alternative reaction mode of the excited singlet state, aldehyde formation is much less frequently observed than decarbonylation. The formation of aldehydes is found in a number of compounds as a minor reaction pathway in competition with the 1,3-acyl shift. [25,26,53]

Ref. [25]

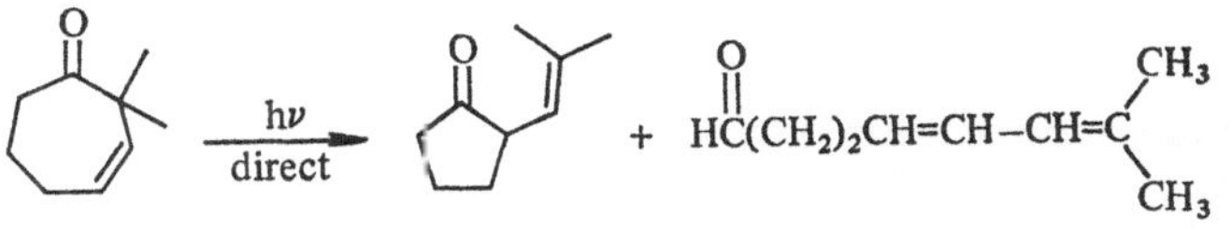

In *127*, aldehyde formation is the only singlet reaction found to occur. [36b] Aldehyde formation from the triplet state has not (yet) been found.

127

Ref. [36b]

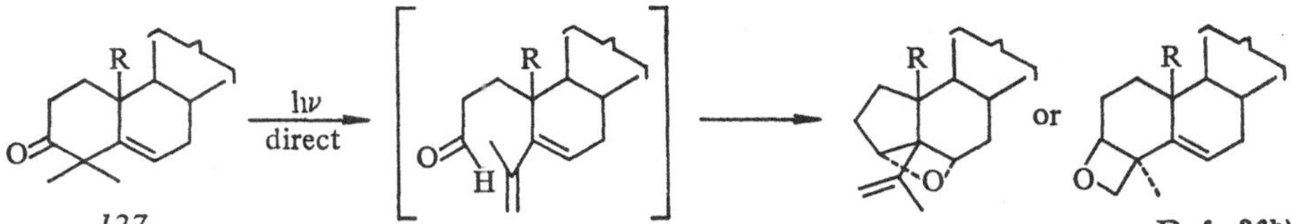

V. Epimerization

Compound *128* gives on acetone sensitized irradiation photoepimerization to *129*, along with a small amount of the products *130* and *131*, which are also the ones obtained on direct irradiation. [79,91]

128 *129* *130* *131*

130 + *131* *132*

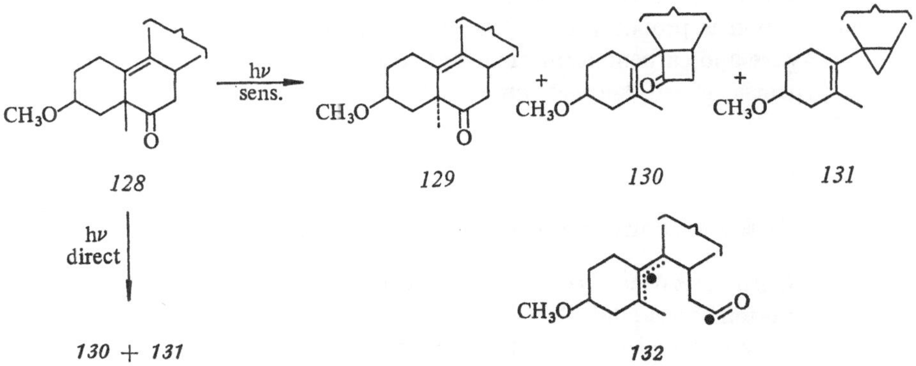

No products of a 1,2-acyl shift are found. The epimerization presumably goes via the diradical *132*. [91] This reaction then gives negative evidence that a (free) diradical formed from α-cleavage is neither the initial intermediate in 1,2-acyl, nor in 1,3-acyl migration. (See Sections I.4 and I.5 for the discussion of the mechanisms). The small degree of photoracemization of *81* under sensitized irradiation possibly provides another example of triplet α-cleavage. [68]

VI. Oxa-carbene Formation

Irradiation of *100* in methanol gives in addition to decarbonylation and 1,3-acyl shift products also the cyclic acetals *133* and *136*. [73] The formation of these compounds is best interpreted as involving the formation of the oxa-carbenes *134* and *135*, followed by addition of solvent. [73]

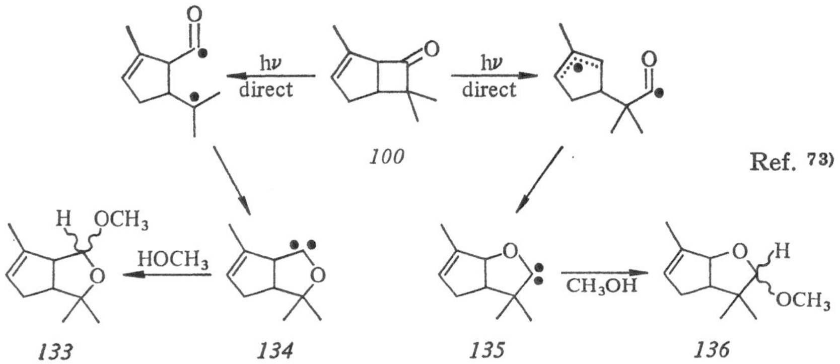

100

Ref. [73]

133   *134*   *135*   *136*

Analogous oxa-carbene intermediates are also indicated in the irradiation of *15b* and *34* in methanol. [20,37] In these cases the first step in the reaction is probably the 1,3-acyl shift to *16b* and *101* followed by α-cleavage and carbene formation. This type of reaction is not restricted to β,γ-unsaturated ketones but is observed also in many saturated ketones. [92]

VII. *Cis-trans* Isomerization and Free Rotor Effect

Cis-trans isomerization of the β,γ-double bond directly from the singlet state has not often been found. [18,48,97] Compound *137* upon direct irradiation displayed an inefficient *cis-trans* isomerization. [18] The isomerization of

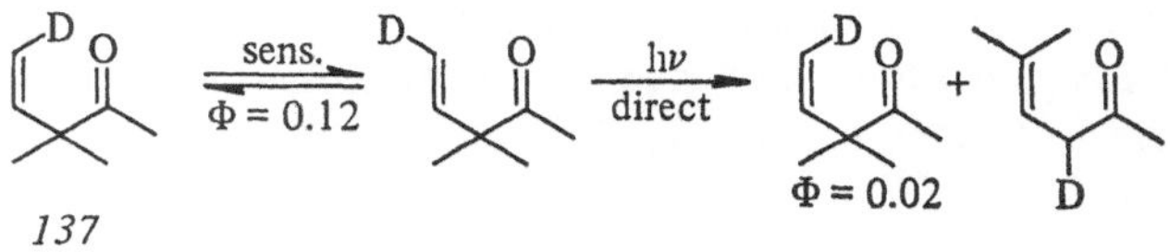

Ref. [18]

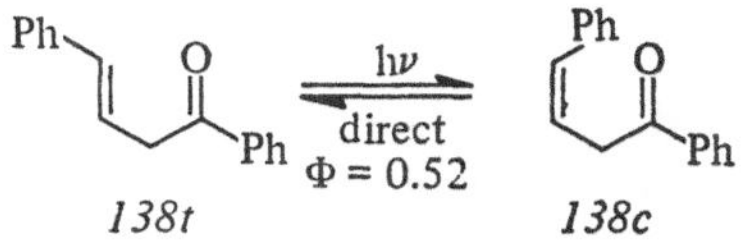

137

Ref. [48]

138t   138c

138 on direct irradiation was efficient and could only be partially quenched by triplet quenchers. [48] This result could be accounted for either by the non-quenchable part of the reaction stemming from the singlet excited state or by a mixing of the n,π* and π,π* triplets.

Unlike the situation in the singlet excited state, *cis-trans*-isomerization is very effective from the triplet state. [9,18,20,48] When the carbon-carbon double bond is not geometrically constrained (*i. e.* the bond is part of an acyclic system or not in a small ring) *cis-trans*-isomerization can become so efficient that it is (almost) the only triplet process observed and the other triplet processes do not compete. This effective process has been called the *free rotor effect* for β,γ-unsaturated ketones in view of the similarity to the concept proposed earlier for flexible di-π-methane systems. [29] Flexible double bonds thus provide an efficient pathway for dissipation of the energy of the triplet state. The free rotor effect is a chemical pathway of intersystem crossing. The effect is nicely borne out

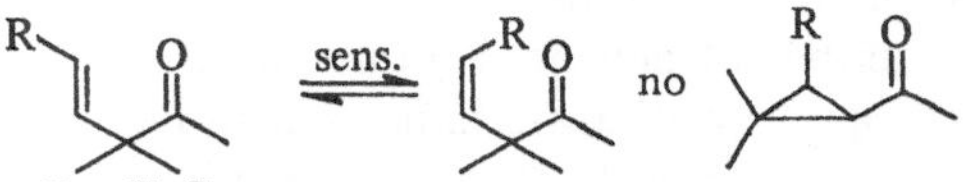

R = H, D

139t   139c

Ref. [18]

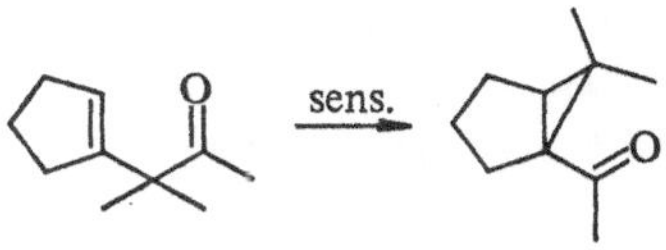

13

Ref. [18]

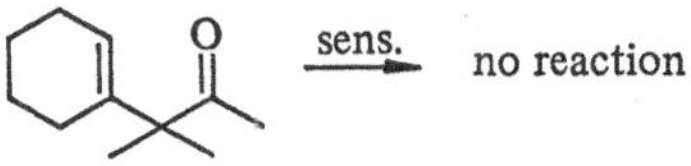

no reaction

140

Ref. [18]

by the study on the closely related compounds *139, 13* and *140*. [18)] Whereas *139* and *140* do not give the oxa-di-π-methane rearrangement, *13* with its C=C bond constrained in a five-membered ring does. The six-membered ring in *140* is apparently big enough to allow twisting of the double bond. *trans*-Cyclohexenes have been repeatedly proposed as intermediates in various photoprocesses but their actual existence has never been established. [98)]

Twisting around the double bond does indeed occur as shown by *139* for R=D. [18)] The concept was first invoked with β,γ-unsaturated ketones for compound *141*; *141a* with R=H seems inert on sensitized irradiation, but *141*b with R=CH₃ shows that *cis, trans*-isomerization occurs. [29)]

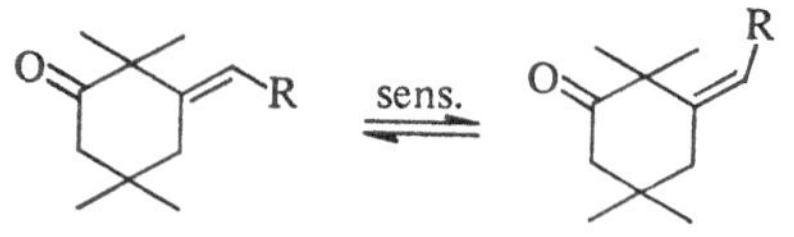

141   Ref. [29)]

a) R = H   b) R = CH₃

Quantum yields have been reported in only a few cases. The sensitized transformation of *139t* to *139c* has a quantum yield of 0.12 [18)] and the triplet portion of the isomerization of *138t* to *138c* a value of 0.35. [48)] With the cyclopentene analog *13* which cannot *cis-trans* isomerize, products are formed with quantum yields of 0.01 and 0.04. In these cases there is an appreciable quantum deficit which indicates there are other deactivating processes occurring. The free rotor effect is not a prohibitive process but only a limiting process. In agreement with this finding is the fact that a number of acyclic β,γ-unsaturated ketones upon sensitization undergo an oxa-di-π-methane rearrangement, albeit inefficiently. It has been found [9,68)] that *11* and the related compound *81* (see Section I.5) on sensitization first establish an efficient *cis,trans* isomer photoequilibrium, followed by an inefficient formation of the cyclopropyl ketone. The sensitized transformation of *86a* and *86b* to *87b* and *87a*, respectively, also occurs very inefficiently. The compounds *142* [99)] and *143* [99,99a)] upon sensitized irradiation undergo no

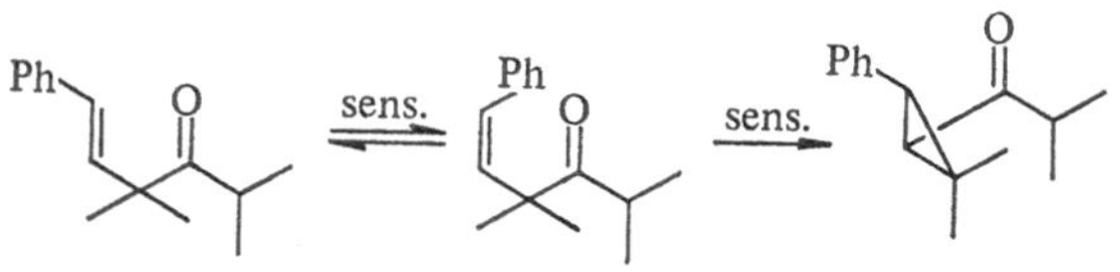

11   Ref. [9)]

structural transformations. This non-reactivity can partially be ascribed to the free rotor effect but the position of placement of the substituent on the double bond also must effect the efficiency of the rearrangement process.

The 1,2-acyl shift will occur when it can compete with the free rotor energy dissipation, the shift can compete when it is more efficient and will be more efficient with increased orbital mixing in the triplet state. The enhancement of the n—π* absorption also reflects the amount of the orbital mixing in the triplet state; thus, in compounds with enhanced n—π* absorption the 1,2-shift can more effectively compete with the energy dissipation along other pathways. This would explain why *11t* ($\varepsilon_{293} = 376$) and *81* ($\varepsilon_{290} = 350$) give an oxa-di-π-methane rearrangement and *142* ($\varepsilon_{291} = 11$) and *143* (*143b* $\varepsilon_{295} = 120$) [99] do not. The other acyclic compound, *86*, where the rearrangement occurs also has somewhat enhanced n—π* absorption. ($\varepsilon_{300} = 102$). [95] The compounds with un-

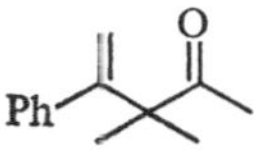

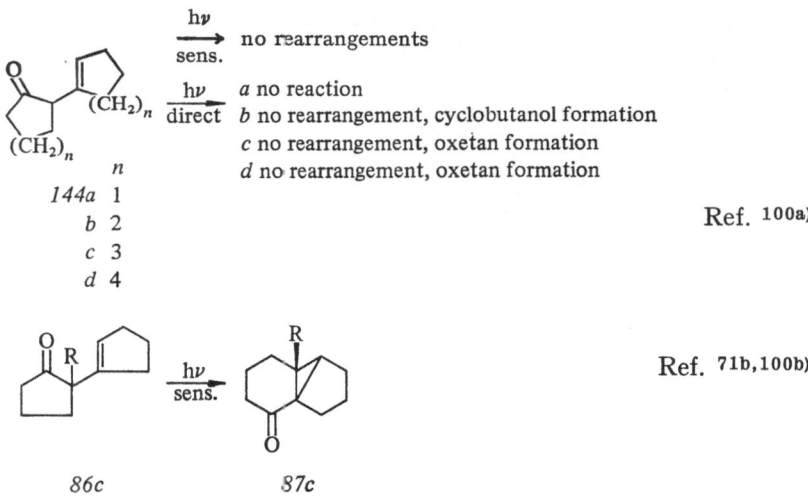

142

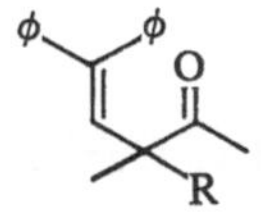

143

R = CH$_3$, C$_2$H$_5$

constrained C=C, which show no rearrangement have either a slight or no n—π* enhancement. Of the 8,7,6 and 5-membered ring compounds *144a–d*, none gives a rearrangement on triplet sensitization. [100]

| | $hν$ → no rearrangements |
| | sens. |

	$hν$	*a* no reaction
	direct	*b* no rearrangement, cyclobutanol formation
		c no rearrangement, oxetan formation
		d no rearrangement, oxetan formation

	n
144a	1
b	2
c	3
d	4

Ref. [100a]

Ref. [71b, 100b]

86c *87c*

R = CH$_3$, CH$_2$CO$_2$Me, CH$_2$COOH

105

The five membered ring analog *86c* with the double bond frozen in the β,γ position underwent an efficient oxa-di-π-methane rearrangement. [71b,100b] Down to the six membered ring compound, the non reactivity can be ascribed to the bond twisting or free rotor effect. In the five membered ring, such motions are restricted and do not dissipate the energy of the triplet, permitting rearrangement. It has been reported [101]

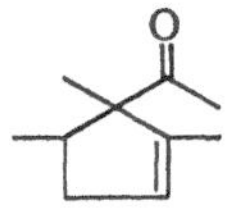

Ref. [101]

145

that *145* gave only less than 5% of an unknown product upon irradiation, a surprising result in view of the rearrangement of *12, 83, 86c*. [17,67, 69,100b] There is no ready explanation for this result.

VIII. Reduction of the Carbon Carbon Double Bond

In a few compounds, photoreduction of the C=C bond effectively competes with the more expected reactions. [24,85,102] In all cases, features are present which make a diminished efficiency of the competing reactions plausible. One example already mentioned is *114*, where reduction is concurrent with an ineffective decarbonylation. [85]

In *146* and *147* reduction takes place to the exclusion of other singlet processes, or if the reduction of *146* and *147* is a triplet reaction by analogy with *114* (the multiplicity of the reduction of *146* and *147* was not studied) *isc* takes precedence over other singlet processes and then reduction over other triplet reactions. [24,102]

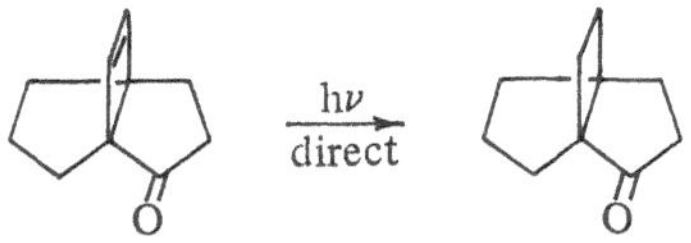

Ref. [102]

146 $\epsilon_{310}=102$

Ref. [24]

147 $\epsilon_{307}=101$

Comparison of the behavior of *146* and *147* with that of their analogs *18* and *20*, which bear two methyl substituents on the C=C bond, shows that the last two compounds display the normal 1,3-acyl shifts and no reduction. The absence of the methyl groups apparently decreases orbital mixing, as borne out by the n—π* absorption coefficients: ε_{311} of *20* = 349 vs. ε_{310} of *146* = 102 and ε_{307} of *18* = 210 vs ε_{307} of *147* = 101. [24, 27, 102] Decreased orbital mixing gives decreased efficiency of the 1,3-acyl shift, thus enabling the reduction (or *isc* followed by reduction) to take over.

IX. Norrish Type II with Cyclobutanol Formation

If a γ-hydrogen is present which is easily available for abstraction, the Norrish type II reaction seems to take precedence over all other possible singlet processes. [18,100,101,103–107]

For example, *148* [104] gives mainly *149* and only a small yield of the 1,3-acyl shift product *150*, while the compound *139* without the methyl group on γ-carbon gives mainly 1,3-acyl shift. [18] In the same way, *140* gives only cyclobutanol formation while the five-ring equivalent *13*

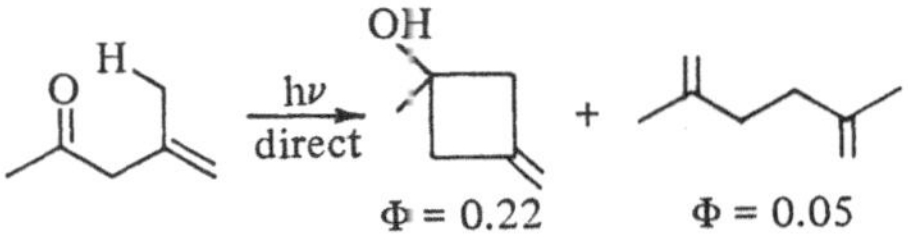

Φ = 0.22   Φ = 0.05   Ref. [103]

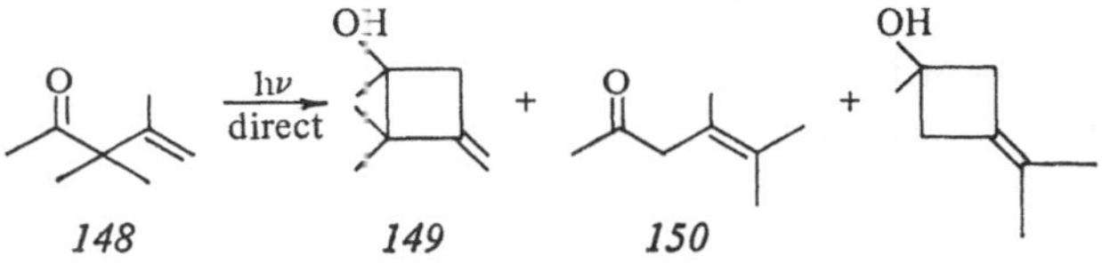

148   *149*   *150*   Ref. [104]

144b   Ref. [100]

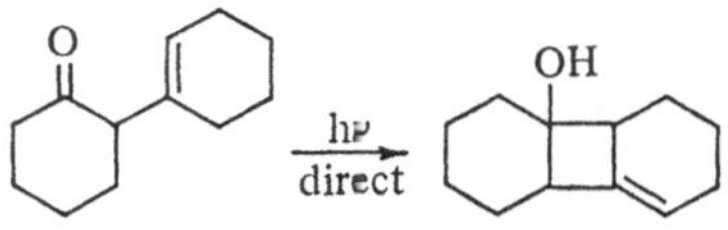

140   Ref. [18]

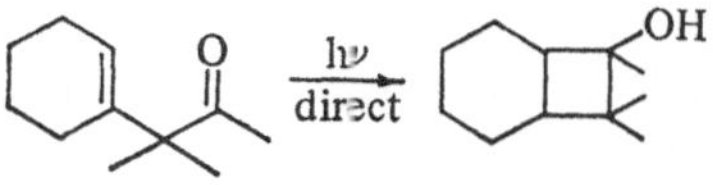

107

gives only 1,3-shift product *14*. [18] Whereas *144b* only gives cyclobutanol formation, the seven- and eight-membered ring equivalents *144c* and *144d* only give oxetan formation. (See Section X). [100] The results obtained by the irradiation of *151* and *153* clearly demonstate the intermediacy of a relatively long-lived 1,4-biradical, such as *152*, in cyclobutanol formation from the enone n,π^* excited singlet state. [107]

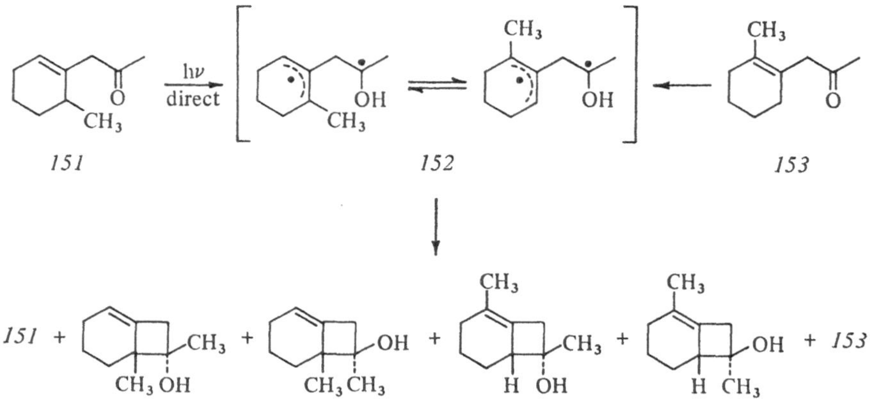

Ref. [107]

X. Cycloaddition Reactions

1. Intramolecular Oxetan Formation

The intramolecular reaction of the ethylenic and carbonyl units of a β,γ-unsaturated ketone to form an oxetan has recently been reported for the cyclooctenyl ketone *144d* which on direct irradiation yields the oxabicyclopentane structure *154*.[100a] The seven-membered ring analog *144c* shows the same reaction, also from the excited singlet state.[100a]

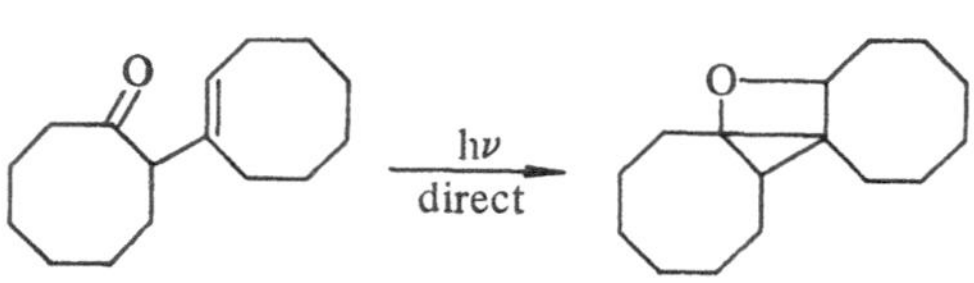

Ref. [100a]

In *144b*, the six-membered analog, however, a Norrish II reaction prevails (Section IX). None of the three compounds shows any 1,3-acyl shift.[100a]

108

Oxetan formation of the carbonyl group with a C=C bond which is an extension of the basic chromophore occurs in *155*.[108] No examples of oxetan formation from the triplet excited state are known.

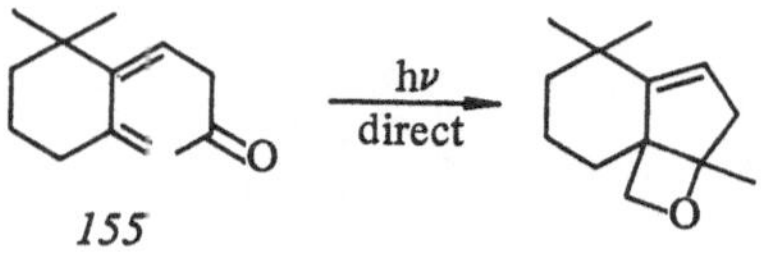

155

Ref. [108]

2. Cyclobutane Formation

Intramolecular cyclobutane formation from the singlet state takes place in suitable systems where an additional carbon-carbon double bond is available.[24,46,56,69,108,109].

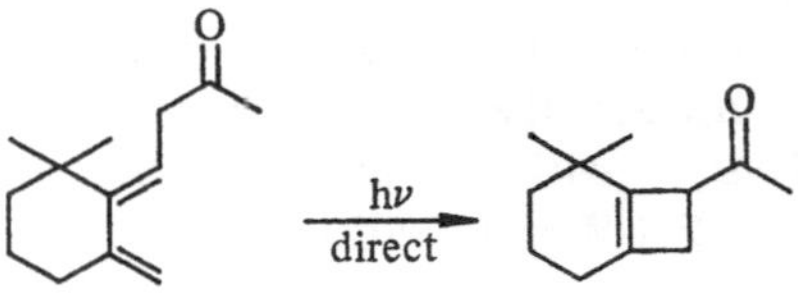

Ref. [108]

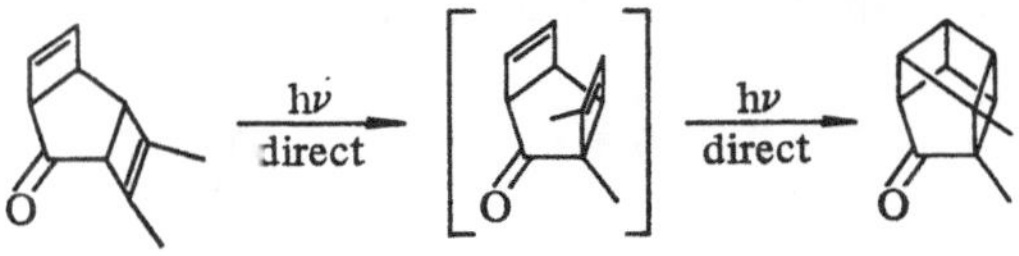

Ref. [24]

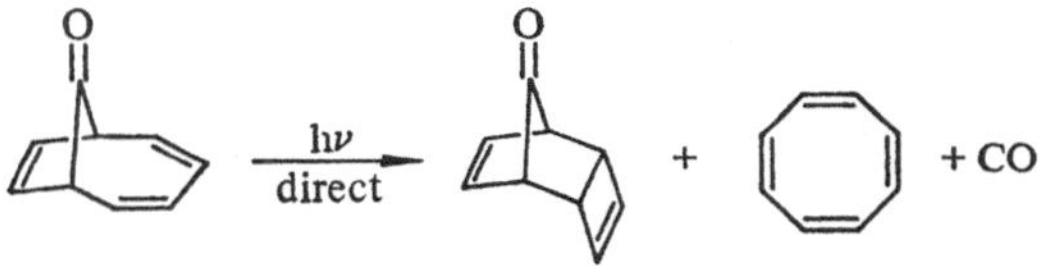

Ref. [46]

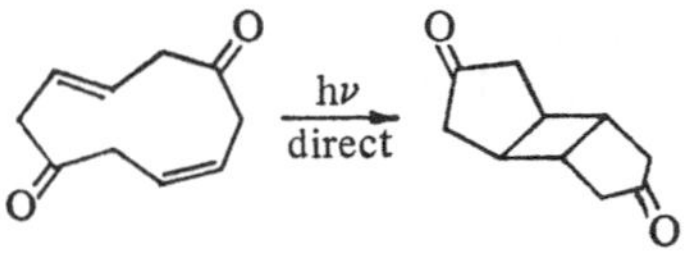

Ref. [109]

There are also a few examples of intramolecular cyclobutane formation from the triplet state.

Compound *98*, which gives efficient decarbonylation from the singlet excited state [Eq. (29)] gives the cyclobutane product *157* on sensitized irradiation.[78,79] The reaction is proposed to go via the photochemical

109

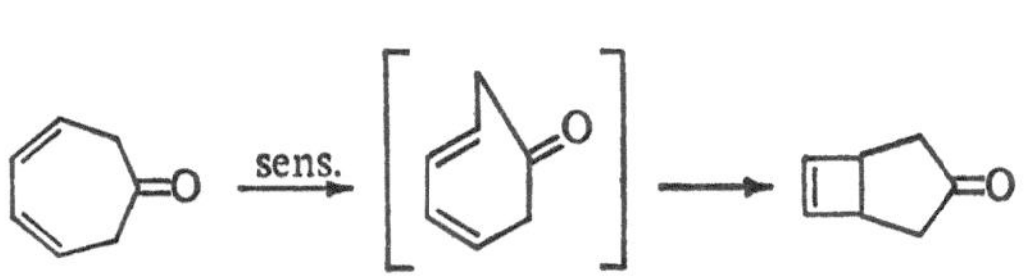

Ref. [78,79]

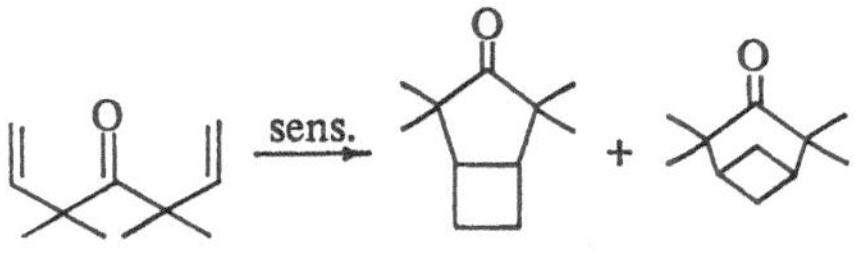

Ref. [80]

formation of the *cis,trans*-diene *156*, which thermally closes to *157*. The proposed intermediate diene could not be trapped.[78,79] The triplet behavior of *60*, the tetramethyl analog of *98*, as well as its singlet behavior, is in sharp contrast to that of *98*.[25] The triplet of *60* gives a 1,2-acyl shift product instead of cyclobutane formation; the methyl groups would sterically inhibit the formation of the *cis-trans*-diene.

The free rotor effect apparently has not much influence on the reactivity of the triplet of *99*.[80] Although no oxa-di-π-methane product is found, efficient ($\phi = 0.19$) [2 + 2]cycloaddition to *158* and *159* takes place.

One example of an intermolecular [2 + 2]cycloaddition has been reported.[102] The multiplicity of this reaction is not known.

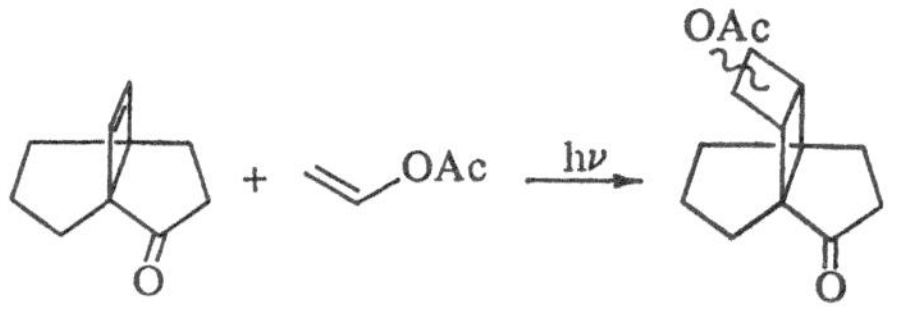

Ref. [102]

Acknowledgement: This study was kindly supported by Grant No. AM-00709, National Institute of Arthritis and Metabolic Diseases, U. S. Public Health Service. The authors also wish to acknowledge support from the Deutsche Forschungsgemeinschaft (Junes Ipaktschi) and from the Senior Fulbright-Hays Program (Gerrit Lodder).

XI. References

[1] Labhart, H., Wagnière, G.: Helv. Chim. Acta *42*, 2219 (1959). — Cookson, R. C., Wariyar, N. S.: J. Chem. Soc. *1956*, 2302.

[2] Moscowitz, A.: Proc. Roy. Soc. (London), Ser. A *297*, 40 (1967). — Cookson, R. C.: Proc. Roy. Soc. (London), Ser. A *297*, 27 (1967). — Murrell, J. N. in: The theory of the electronic spectra of organic molecules, p. 167, New York,

J. Wiley & Sons 1963. — Winstein, S., de Vries, L., Orloski, R.: J. Am. Chem. Soc. *83*, 2020 (1961). — Kosower, E. M., Closson, W. D., Goering, H. L., Gross, J. C.: J. Am. Chem. Soc. *83*, 2013 (1961). — Cookson, R. C., Hill, R. R., Hudec, J.: Chem. Ind. *1961*, 589.

[3] Höhn, E. G., Weigang, Jr., O. E.: J. Chem. Phys. *48*, 1127 (1968). — Bays, D. E., Cookson, R. C., MacKenzie, S.: J. Chem. Soc. B *1967*, 215. — Mason, S. F.: Quart. Rev. *17*, 20 (1963). — Moskovitz, A., Mislow, K., Glass, M. A. W., Djerassi, C.: J. Am. Chem. Soc. *84*, 1945 (1962).

[4] Chadwick, D., Frost, D. C., Weiler, L.: J. Am. Chem. Soc. *93*, 4320, 4962 (1971).

[5] Marsh, G., Kearns, D. R., Schaffner, K.: J. Am. Chem. Soc. *93*, 3129 (1971).

[6] Hancock, K. G., Grider, R. O.: Chem. Commun. *1972*, 580.

[7] Houk, K. N., Northington, D. J., Duke, Jr., R. E.: J. Am. Chem. Soc. *94*, 6233 (1972).

[8] Zimmerman, H. E., Amick, D. R.: J. Am. Chem. Soc. *95*, 3977 (1973). — Zimmermann, H. E., Marianc, P. S.: J. Am. Chem. Soc. *91*, 1718 (1969).

[9] Dauben, W. G., Kellogg, M. S., Seeman, J. I., Spitzer, W. A.: J. Am. Chem. Soc. *92*, 1786 (1970).

[10] Büchi, G., Burgess, E. M.: J. Am. Chem. Soc. *82*, 4333 (1960).

[11] Ipaktschi, J.: Chem. Ber. *105*, 1996 (1972). — Ipaktschi, J.: Tetrahedron Letters *1970*, 3179.

[12] Schuster, D. I., Sussman, D. H.: Tetrahedron Letters *1970*, 1661.

[13] Ipaktschi, J.: Tetrahedron Letters *1969*, 215.

[14] Givens, R. S., Oettle, W. F.: Chem. Commun. *1969*, 1164.

[15] Givens, R. S., Oettle, W. F., Coffin, R. L., Carlson, R. G.: J. Am. Chem. Soc. *93*, 3957 (1971).

[16] Givens, R. S., Oettle, W. F.: J. Am. Chem. Soc. *93*, 3963 (1971).

[17] Baggiolini, E., Schaffner, K., Jeger, O.: Chem. Commun. *1969*, 1103.

[18] Engel, P. S., Schexnayder, M. A.: J. Am. Chem. Soc. *94*, 9252 (1972).

[19] a) Schuster, D. I., Axelrod, M., Auerbach, J.: Tetrahedron Letters *1963*, 1911; b) Bays, D. E., Cookson, R. C.: J. Chem. Soc. B *1967*, 226.

[20] Ipaktschi, J.: Chem. Ber. *105*, 1840 (1972). — Ipaktschi, J.: Tetrahedron Letters *1969*, 2153.

[21] Carlson, R. G., Bateman, J. H.: Tetrahedron Letters *1967*, 4151.

[22] Erman, W. F., Kretschmar, H. C.: J. Am. Chem. Soc. *89*, 3842 (1967).

[23] Meinwald, J., Janse-van Vuuren, P.: Chem. Commun. *1971*, 1460.

[24] Cargill, R. L., King, T. Y., Sears, A. B., Willcott, M. R.: J. Org. Chem. *36*, 1423 (1971); see also Yano, K.: Tetrahedron Letters *1974*, 1861.

[25] Paquette, L. A., Eizember, R. F., Cox, O.: J. Am. Chem. Soc. *90*, 5153 (1968).

[26] Paquette, L. A., Eizember, R. F.: J. Am. Chem. Soc. *89*, 6205 (1967). — Crandall, J. K., Arrington, J. P., Hen, J.: J. Am. Chem. Soc. *89*, 6208 (1967).

[27] Cargill, R. L., Beckham, M. E., Sibert, A. E., Dorn, J.: J. Org. Chem. *30*, 3647 (1965).

[28] Furutachi, N., Nakadaira, Y., Nakanishi, K.: J. Am. Chem. Soc. *91*, 1028 (1969). — Hayashi, J., Furutachi, N., Nakadaira, Y., Nakanishi, K.: Tetrahedron Letters *1969*, 4589. — Furutachi, N., Hayashi, J., Sato, H., Nakanishi, K.: Tetrahedron Letters *1972*, 1061.

[29] Hancock, K. G., Grider, R. O.: Tetrahedron Letters *1972*, 1367; Tetrahedron Letters *1971*, 4281; J. Am. Chem. Soc. *96*, 1158 (1974).

[30] Pfenninger, E., Poel, D. E., Berse, C., Wehrli, H., Schaffner, K., Jeger, O.: Helv. Chim. Acta *51*, 772 (1968).

[31] Ipaktschi, J.: unpublished results.

111

32) Criegee, R., Furrer, H.: Chem. Ber. *97*, 2949 (1964). — Eaton, P. E.: Tetrahedron Letters *1964*, 3695.

33) Murata, I., Sugihara, Y.: Chem. Letters *1972*, 625.

34) Hart, H., Murray, Jr., R. K., Appleyard, G. D.: Tetrahedron Letters *1969*, 4785.

35) Hixson, S. S., Mariano, P. S., Zimmerman, H. E.: Chem. Rev. *1973*, 531.

36) a) Kojima, K., Sakai, K., Tanabe, K.: Tetrahedron Letters *1969*, 1925;
   b) Kojima, K., Sakai, K., Tanabe, K.: Tetrahedron Letters *1969*, 3399.

37) Scharf, H. D., Kusters, W.: Chem. Ber. *104*, 3016 (1971).

38) Knott, P. A., Mellor, J. M.: Tetrahedron Letters *1970*, 1829.

39) Knott, P. A., Mellor, J. M.: J. Chem. Soc., Perkin Trans. I, 1030 (1972); see on the other hand Mellor, J. M., Webb, C. F.: J. Chem. Soc., Perkin Trans. I, *1972*, 211.

40) Domb, S., Bozzato, G., Saboz, J. A., Schaffner, K.: Helv. Chim. Acta *52*, 2436 (1969).

41) Domb, S., Schaffner, K.: Helv. Chem. Acta *53*, 677 (1970).

42) Swenton, J. S., Madigan, D. M.: Tetrahedron *28*, 2703 (1972).

43) Jones, D. W., Kneen, G.: Chem. Commun. *1972*, 1038.

44) Hart, H., Murray, Jr., R. K.: Tetrahedron Letters *1969*, 379. — Murray, Jr., R. K., Hart, H.: Tetrahedron Letters *1968*, 4995.

45) Goldschmidt, Z., Kende, A. S.: Tetrahedron Letters *1971*, 4625. — Kende, A. S., Goldschmidt, Z.: Tetrahedron Letters *1970*, 783. — Chapman, O. L., Kane, M., Lassila, J. D., Loeschen, R. L., Wright, H. E.: J. Am. Chem. Soc. *91*, 6856 (1969). — Kende, A. S., Goldschmidt, Z., Izzo, P. T.: J. Am. Chem. Soc. *91*, 6858 (1969). — Ciabattoni, J., Crowley, J. E., Kende, A. S.: J. Am. Chem. Soc. *89*, 2778 (1967).

46) Kurabayashi, K., Mukai, T.: Tetrahedron Letters *1972*, 1049. — Antkowiak, T. A., Sanders, D. C., Trimitsis, G. B., Press, J. B., Shechter, H.: J. Am. Chem. Soc. *94*, 5366 (1972). — Paquette, L. A., Meisinger, R. H., Wingard, R. E.: J. Am. Chem. Soc. *94*, 2155 (1972).

47) Tenney, L. P., Boykin, Jr., D. W., Lutz, R. E.: J. Am. Chem. Soc. *88*, 1835 (1966).

48) Cowan, D. O., Baum, A. A.: J. Am. Chem. Soc. *93*, 1153 (1971).

49) Williams, J. R., Ziffer, H.: Chem. Commun. *1967*, 194.

50) Williams, J. R., Ziffer, H.: Tetrahedron *24*, 6725 (1968). — Williams, J. R., Ziffer, H.: Chem. Commun. *1967*, 468.

51) Engel, P. S., Schexnayder, M. A., Ziffer, H., Seeman, J. I.: J. Am. Chem. Soc. *96*, 924 (1974).

52) Williams, J. R., Sarkisian, G. M.: Chem. Commun. *1971* 1564.

53) Sato, H., Furutachi, N., Nakanishi, K.: J. Am. Chem. Soc. *94*, 2150 (1972).

54) Sato, H., Nakanishi, K., Hayashi, J., Nakadaira, Y.: Tetrahedron *29*, 275 (1973).

55) Houk, K. N., Northington, D. J.: J. Am. Chem. Soc. *94*, 1387 (1972).

56) Mukai, T., Akasaki, Y., Hagiwara, T.: J. Am. Chem. Soc. *94*, 675 (1972). — Akasaki, Y., Mukai, T.: Tetrahedron Letters *1972*, 1985.

57) Griffiths, J., Hart, H.: J. Am. Chem. Soc. *90*, 5296 (1968).

58) Schuster, D. I., Underwood, G. R., Knudsen, T. P.: J. Am. Chem. Soc. *93*, 4304(1971).

59) Carlson, R. G., Henton, D. A.: Chem. Commun. *1969*, 674.

60) Woodward, R. B., Hoffmann, R.: Angew. Chem. *81*, 797 (1969); Angew. Chem. Intern. Ed. Engl. *8*, 781 (1969).

61) Gozenbach, H. U., Schaffner, K., Blank, B., Fischer, H.: Helv. Chim. Acta *56*, 1741 (1973).

62) Strausz, O. P., Kozak, P. J., Woodall, G. N. C., Sherwood, A. G., Gunning, H. E.: Can. J. Chem. *46*, 1317 (1968).

63) Footnote 17 in Ref. 9). See also Zimmerman, H. E., Flechtner, T. W.: J. Am. Chem. Soc. *92*, 6931 (1970). — Zimmerman, H. E.: Hancock, K. G., Licke, G. C.: J. Am. Chem. Soc. *90*, 4892 (1968).

64) Herndon, W. C., Giles, W. B.: Mol. Photochem. *2*, 277 (1970). — Michl, J.: Mol. Photochem., *4*, 257 (1972).

65) Fukui, K.: Accounts Chem. Res. *4*, 57 (1971).

66) Ipaktschi, J.: Habilitationsschrift. Universität Heidelberg 1972.

67) Gonzenbach, H. U., Ph. D. Thesis,: Eidgenössische Technische Hochschule Zürich 1973.

68) Dauben, W. G., Lodder, G.: unpublished results.

69) Schaffner, K.: Pure Appl. Chem. *33*, 329 (1973).

70) Dauben, W. G., Welch, W. M.: Tetrahedron Letters *1971*, 4531. — Zimmerman, H. E., Hancock, K. G.: J. Am. Chem. Soc. *90*, 4892, (1968).

71) a) Seeman, J. I., Ziffer, H.: Tetrahedron Letters *1973*, 4409, 4413.;
b) Coffin, R. L., Carlson, R. G., Givens, R. S. in: Abst. of 167th Meeting of the American Chemical Society, Los Angeles, CA., April 1—5, p. ORGN-80.

72) a) Plank, D. A., Floyd, J. C.: Tetrahedron Letters 1971, 4811. — Matsuura, T., Ogura, K.: J. Am. Chem. Soc. *89*, 3850 (1967);
b) Dauben, W. G., Welch, W. M.: Tetrahedron Letters 1971, 4531.

73) Erman, W.: J. Am. Chem. Soc. *89*, 3828 (1967).

74) Chambers, R. J., Marples, B. A.: Tetrahedron Letters 1971, 3751.

75) Dowd, P., Gold, A., Sachev, K.: J. Am. Chem. Soc. *92*, 5724, 5725 (1970). — Dowd, P., Senqupta, G., Sachdev, K.: J. Am. Chem. Soc. *92*, 5726 (1970).

76) Mukai, T., Kurabayashi, K.: J. Am. Chem. Soc. *92*, 4493 (1970).

77) Chapman, O. L., Bordon, G. W.: J. Org. Chem. *26*, 4185 (1961). — Chapman, O. L., Pasto, D. J., Borden, G. W., Griswold, A. A.: J. Am. Chem. Soc. *84*, 1220 (1962).

78) Schuster, D. I., Sckolnick, B. R., Lee, F. T. H.: J. Am. Chem. Soc. *90*, 1300 (1968).

79) Schuster, D. I., Blythin, D. J.: J. Org. Chem. *35*, 3190 (1970).

80) Engel, P. S., Schexnayder, M. A.: J. Am. Chem. Soc. *94*, 4357 (1972).

81) Schenck, G. O., Steinmetz, R.: Chem. Ber. *96*, 520 (1963).

82) Baggiolini, E., Hamlow, H. P., Schaffner, K.: J. Am. Chem. Soc. *92*, 4906 (1970).

83) Schuster, D. I., Lee, F. T. H., Padwa, A., Gassman, P. G.: J. Org. Chem. *30*, 2262 (1965). — Fuchs, B.: J. Chem. Soc. C *1968*, 68. — Fuchs, B., Yankelevich, S.: Israel J. Chem. *1968*, 511. — Fuchs, B.: Israel J. Chem. *1968*, 517. — Wilson, W. S., Warrener, R. N.: Tetrahedron Letters *1970*, 5203. — Fuchs, B., Scharf, G.: J. Chem. Soc. Chem. Commun. *1974*, 226.

84) Starr, J. E., Eastman, R. H.: J. Org. Chem. *31*, 1393 (1966).

85) Engel, P. S., Ziffer, H.: Tetrahedron Letters *1969*, 5181.

86) Quinkert, G., Palmowski, J., Lorenz, H. P., Wiersdorf, W. W., Finke, M.: 169 Angew. Chem. *83*, 210 (1971); Angew. Chem. Intern. Ed. Engl. *10*, 196 (1971).

86a) Darling, T. R., Pouliquen J., Turro, N. J.: J. Am. Chem. Soc. *96*, 1247 (1974).

87) Bellamy, A. J., Crilly, W.: J. Chem. Soc., Perkin II *1973*, 112. — Bellamy, A. J., Crilly, W.: J. Chem. Soc., Perkin II *1972*, 395. — Baldwin, J. E., Krueger, S. M.: J. Am. Chem. Soc. *91*, 2396 (1969).

88) Bastiani, R. J., Hart, H.: J. Org. Chem. *37*, 2830 (1972). — Hart, H., Collins, P. M., Waring, A. J.: J. Am. Chem. Soc. *88*, 1005 (1966).
89) Quinkert, G.: Pure Appl. Chem. *33*, 285 (1973).
90) Chapman, O. L., Kane, M., Lassila, J. D., Loeschen, R. L., Wright, H. E.: J. Am. Chem. Soc. *91*, 6856 (1969). — Chapman, O. L., Lassila, J. D.: J. Am. Chem. Soc. *90*, 2449 (1968). — Dauben, W. G., Koch, K., Smith, S. L., Chapman, O. L.: J. Am. Chem. Soc. *85*, 2616 (1963). — Kende, A. S., Goldschmidt, Z., Izzo, P. T.: J. Am. Chem. Soc. *91*, 6858 (1969).
91) Chambers, R. J., Marpless, B. A.: Tetrahedron Letters *1971*, 3747.
92) Yates, P., Kilmurry, L.: Tetrahedron Letters *1964*, 1739. — Yates, P., Kilmurry, L.: J. Am. Chem. Soc. *88*, 1563 (1966).
93) Gloor, I., Schaffner, K.: Jeger, O.: Helv. Chim. Acta *54*, 1864 (1971).
94) Takeda, K., Horibe, I., Minato, H.: Chem. Commun. *1971*, 87.
95) Fischer, M., Zeeh, B.: Chem. Ber. *101*, 2360 (1968).
96) Anet, F. A. L., Mullis, D. P.: Tetrahedron Letters *1969*, 737.
97) Morrison, H.: Tetrahedron Letters *1964*, 3653.
98) Marshall, J. A.: Science *170*, 137 (1970). — Kropp, P. J., Kraus, H. J.: J. Am. Chem. Soc. *89*, 5199 (1967).
99) Dauben, W. G., Kellogg, M. S., Seeman, J., Spitzer, A.: unpublished results.
99a) Pratt, A. C.: J. Chem. Soc. Perkin I, 2496 (1973).
100) a) Cookson, R. C., Rogers, N. R.: Chem. Commun. *1972*, 809;
b) Carlson, R. G., Coffin, R. L., Cox, W. W., Givens, R. S.: Chem. Commun. *1973*, 501.
101) Conia, J. M., Bertolussi, M.: Bull. Soc. Chim. France *1972*, 3402.
102) Cargill, R. L., Damewood, J. R., Cooper, M. M.: J. Am. Chem. Soc. *88*, 1330 (1966).
103) Yang, N. C., Thap, D. M.: Tetrahedron Letters *1966*, 3671.
104) Kiefer, E. F., Carlson, D. A.: Tetrahedron Letters *1967*, 1617.
105) Cookson, R. C., Hudec, J., Szabo, A., Usher, G. E.: Tetrahedron *24*, 4353 (1968).
106) Matsui, T., Komatsu, A., Moroe, T.: Bull. Chem. Soc. Japan *40*, 2204 (1967).
107) Dalton, J. C., Chan, H. F.: J. Am. Chem. Soc. *95*, 4085 (1973).
108) Van Wageningen, A., Cerfontain, H.: Tetrahedron Letters *1972*, 3679.
109) Shani, A.: Tetrahedron Letters *1968*, 5175. — Scheffer, J. R., Lungle, M. L.: Tetrahedron Letters *1969*, 845. — Stankarb, J. W., Conrow, K.: Tetrahedron Letters *1969*, 2395.

Received June 25, 1974

Protein Triplet States

Professor August H. Maki*

Department of Chemistry, University of California, Riverside, California, USA

Dr Joseph A. Zuclich

Technology Incorporated, Life Sciences Division, San Antonio, **Texas,** USA

Contents

* Present address: Department Chemistry, University of California, Davis, California, USA

1. Introduction

1.1. Protein Structure and Function

The basic building blocks of all naturally occurring proteins are the amino acids which have the general Zwitterion structure

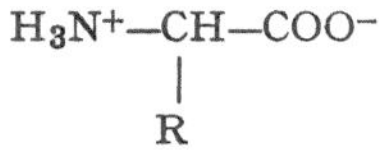

$$H_3N^+-CH-COO^-$$
$$|$$
$$R$$

The amino acids are linked together by peptide bonds (CO—NH) formed by the elimination of water between the amino and carboxyl groups resulting in chains of the form

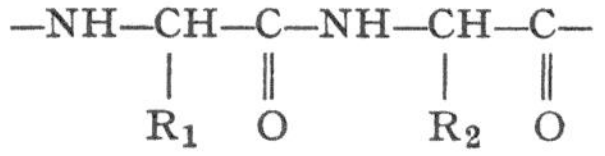

$$-NH-CH-C-NH-CH-C-$$
$$\quad | \quad \| \qquad | \quad \|$$
$$\quad R_1 \quad O \qquad R_2 \quad O$$

Any such chain, whether or not possessed of a natural biological function, may be referred to in the literature as a peptide, polypeptide or oligopeptide. The class of naturally occurring amino acid polymers, called proteins, is found in all organic matter forming integral components of every part of the cell (including the cell membrane, the cytoplasm, the cellular organelles and the chromosomal material) as well as forming the basis for most intercellular structural material. In living matter proteins perform such diverse functions as catalyzing and regulating a great variety of chemical processes (enzymes), endowing organisms with the capacity of motion (contractile proteins), protecting the organism from attack by foreign agents (antibodies) and transferring essential factors from points of ingestion to the places where they are biochemically utilized (transport proteins).

Considering the staggering variety of life forms and the many functions of protein in each life form it is perhaps surprising that no more than about twenty naturally occurring amino acids are found in all naturally occurring proteins. The countless varieties of protein are distinguished from one another by the sequence of amino acids along the peptide chain (primary structure) and, more importantly as far as function is concerned, by the overall three dimensional structure (secondary, tertiary, and quaternary structure) that the protein assumes in its natural medium.

1.2. Early Observation of Protein Phosphorescence

The observation of radiation-induced long-lived luminescence in living material dates back at least until the 18th Century when Becconi [1] reported the observation that a person in a dark room could detect an afterglow from his own skin immediately after that portion of the skin had been exposed to sunlight. It was not until the 1930's, however, that more extensive observations were made on this phenomenon [2,3]. Giese and Leighton [4] in an article entitled "Phosphorescence of Cells and Cell Products" showed that this afterglow, having a lifetime of several seconds or longer, is associated with tissue or tissue products containing very high protein density. This same phenomenon can be observed following UV irradiation of protein powders [5] and solutions [6,7] at room temperature.

Although considerable evidence [8] is presented for involvement of protein triplet states in this phenomenon of afterglow, the mechanism is not that usually associated with the production of phosphorescence. Rather, in the above instances the UV irradiation causes photochemical damage resulting in the production of free radicals [9,10]. The afterglow is the result of luminescence from chemically excited states formed upon radical recombinations.

Phosphorescence from optically excited protein triplet states, however, is readily observable at low temperatures. Debye and Edwards [11] first reported the observation of a brilliant blue phosphorescence from many proteins at liquid nitrogen temperature. Although the evidence presented is scant, they correctly identified the source of this emission as the aromatic amino acid residues of the proteins studied. Subsequent work by Konev [8] and others [12,13] confirmed this by extensive comparisons of the emission and excitation spectra as well as the phosphorescence lifetimes of various proteins and the aromatic amino acids. It is interesting to observe as a historical note that for many years a hypothesis persisted in the literature that protein luminescence was a property of the protein as a whole rather than originating from distinct sites within the protein. It was postulated that electron delocalization existed along the peptide backbone resulting in exciton bands capable of luminescence [14-16].

Debye and Edwards [11] noted two components in the decay of low temperature protein phosphorescence. The greater part of the decay was exponential in character having lifetimes on the order of several seconds. A weak, and much longer-lived component, however, was reported to have the same emission spectrum but a non-exponential decay. Debye and Edwards claimed that this emission was a result of radical recombination following photoejection of an aromatic amino acid electron into the surrounding medium. They showed that this decay followed an inverse power function of time in accordance with the theory for a diffusion controlled recombination process [17].

It should be pointed out before concluding this section, that although observed phosphorescence emission from native proteins is almost always characteristic of emission from the aromatic amino acids, a number of proteins have other unsaturated molecules incorporated into their native structure. Molecules such as heme, flavins and carotenoids when complexed with protein contribute to the near UV and visible absorption of the complex and may be involved in energy transfer with aromatic amino acids. Further, triplet probes can often be bound to specific sites in a protein (e.g. inhibitor molecules bound to the catalytic site of an enzyme) to study such interesting problems as interactions of the probe with metal ions [18] (which may be present in the catalytic site) or with the aromatic amino acids [19,20]. Since such probes are not native to the protein they are not considered further in this report.

Finally, a question which always must be dealt with when discussing observations made on frozen solution of biological macromolecules is whether or not the process of lowering the temperature and freezing the solution has any effect on the molecular conformation or other properties. While a definitive answer cannot be given, a fair amount of data has accumulated to suggest that no untoward effects are encountered in frozen protein solutions. NMR [21] and circular dichroism studies [22] do not indicate any significant changes in conformation upon freezing aqueous solutions of proteins and polypeptides. Laser Raman studies [23] similarly fail to detect any major changes between protein conformation in solution and in the crystalline state. Most striking, however, is the ability of enzymes to function at very low temperatures in dimethyl formamide-H_2O, and ethylene glycol-H_2O solutions [24] despite the usual very high sensitivity of enzyme-substrate interactions to deviation from natural conditions.

2. Aromatic Amino Acids

Of the twenty amino acids found in proteins, three may be classified as aromatic in character and account for virtually all of the protein absorption of ultraviolet light above 250 nm and for all of the observed luminescence. The molecular structures of the aromatic amino acids are shown in Fig. 1. The parent aromatic molecules of tryptophan, tyrosine and phenylalanine are, respectively, indole, phenol and benzene. The parents with methyl substituents in place of the amino acid side chains have the common names skatole, p-cresol and toluene, respectively.

The UV absorption spectra of the aromatic amino acids [25] are compared in Fig. 2. While none of the aliphatic amino acid monomers has any appreciable absorption in this wavelength region, disulfide bridges which are commonly formed between two cysteine residues of a peptide chain contribute an absorption band at 245 nm with a maximum molar extinction coefficient of 350. The peptide bond itself absorbs quite strongly in the far UV and along with the aliphatic amino acids begins to contribute significantly to protein absorption at wavelengths below 220 nm.

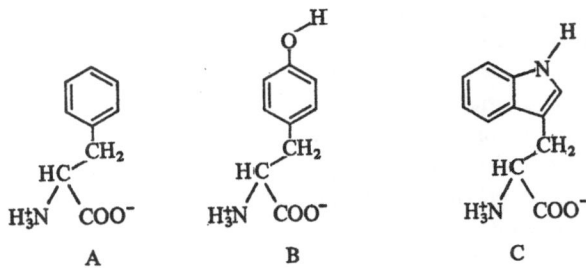

Fig. 1. Structures of the aromatic amino acids $(3 \leq pH \leq 9)$. (A) phenylalanine; (B) tyrosine; (C) tryptophan

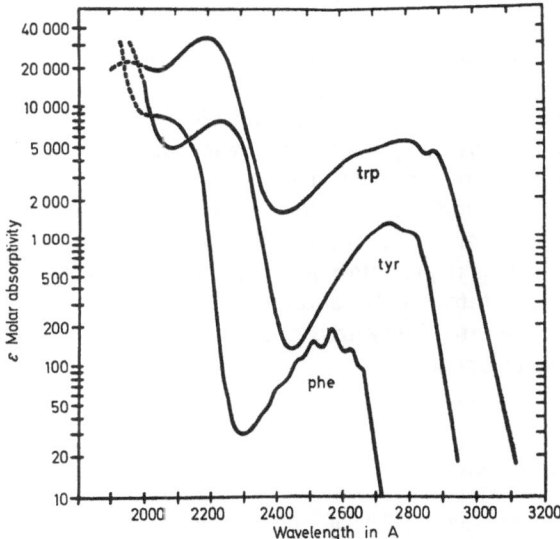

Fig. 2. Ultraviolet absorption spectra of tryptophan, tyrosine, and phenylalanine at pH 6 (from Wetlaufer [25]).
[From Weinryb, Steiner: Excited states of proteins and nucleic acids, pp. 277—318. Fig. 1, p. 279. (1971).]

119

As may be seen from Fig. 2., the absorption of phenylalanine is negligible compared with tyrosine and tryptophan. In addition, as will be discussed in later sections, there is a likelihood of energy transfer from an excited phenylalanine residue to any neighboring tyrosine or tryptophan group in a protein molecule. For these reasons unless there is a complete absence of both tyrosine and tryptophan in a given protein no phenylalanine contribution to the protein luminescence is found. For the same reasons (but to a lesser extent) the tryptophan contribution tends to overshadow the tyrosine luminescence of any protein which contains both of these amino acids. Since tryptophan is a common enough amino acid to be present to varying extents in a majority of proteins the protein luminescence is often found to be primarily due to tryptophan with at most a small contribution from tyrosine luminescence. When tryptophan is not present, the protein luminescence generally is found to be due solely to tyrosine. Thus, as might be expected, studies of the luminescence properties of tryptophan are much more prevalent in the literature than those on tyrosine whereas phenylalanine has been virtually ignored.

As may be inferred from the preceding discussion, "protein triplet states" are in effect equated with the phosphorescent triplet states of the aromatic amino acid residues found in the proteins. There are, in general, no ground state triplets associated with the protein and aside from the aromatic residues no other protein components which can be efficiently excited to a triplet state or have long enough triplet lifetimes for any appreciable triplet populations to accumulate. The aromatic amino acid triplet states, however, can be efficiently excited indirectly by intersystem crossing processes following photon absorption into the excited singlet state manifold. Direct optical excitation to the triplet state is highly forbidden as reflected by the long triplet lifetimes. At low temperatures the aromatic amino acids phosphoresce brightly and because of their long triplet lifetimes high steady state populations of the phosphorescent state can be obtained. Therefore, the aromatic amino acid triplet states are highly amenable to studies by optical spectroscopy and magnetic resonance.

2.1. Phosphorescence

Phosphorescence spectra [26] of the aromatic amino acids in neutral aqueous glucose-containing glasses at liquid nitrogen temperatures are shown in Fig. 3. Since the populating of these phosphorescent triplet states generally proceeds through intersystem crossing from the fluorescent singlet states the fluorescence spectra [27] are shown for reference

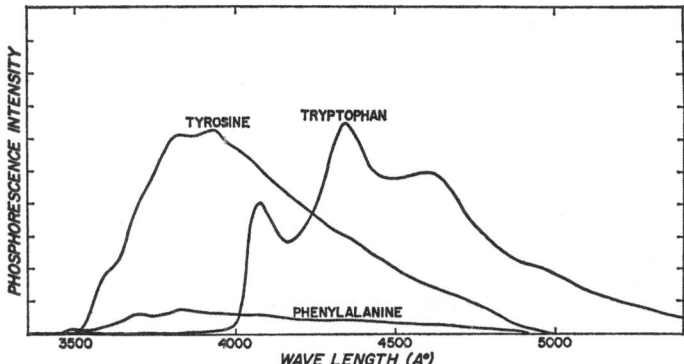

Fig. 3. Phosphorescence emission of tryptophan, tyrosine, and phenylalanine at 10^{-3} M concentration in 0.5% glucose at 77 °K with identical amplification and optics (from Truong et al. [26]).

[From Truong, Bersohn, Brumer, Luk, Tao: J. Biol. Chem. 242, 2979 (1967), Fig. 1.]

in Fig. 4 and pertinent singlet state data are included in the discussion below. For extensive reviews of fluorescence properties of the aromatic

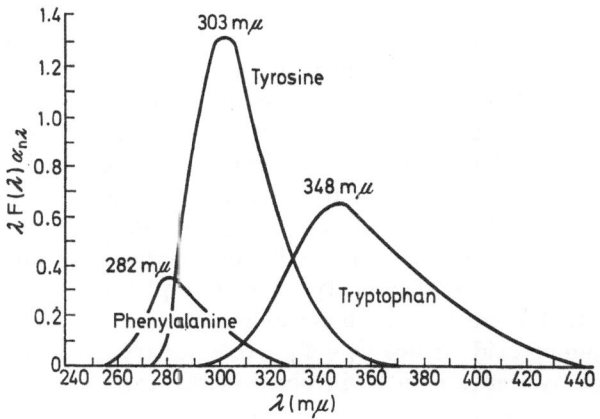

Fig. 4. Corrected fluorescence spectra of the aromatic amino acids in water at 293 °K (from Teale and Weber [27]).

[From Weinryb, Steiner: Excited states of proteins and nucleic acids, pp. 277—318. Fig. 2, p. 292. (1971).]

amino acids and proteins, the reader is referred to three recent texts [8,28a,b]. Absorption and luminescence data are summarized in Table 1;

Table 1. Absorption and luminescence properties of the aromatic amino acids [1]

	Tryptophan		Tyrosine		Phenylalanine		Ref.
Absorption maxima (nm)	280	218	275	222	258	205	[29]
Molar extinction coefficient at maximum	6500	27,000	1290	8000	200	8,500	[29]
Fluorescence maximum (nm)	330		300		290		[30]
Fluorescence quantum yield (Φ_f)	0.72		0.47		0.41		[31]
Phosphorescence maximum (nm)	435		395		385		[30]
Phosphorescence quantum yield (Φ_p)	0.17		0.53		0.59		[31]
Phosphorescence lifetime (sec)	5.8		2.7		5.5		[30]
	6.65		2.90		7.70		[28b]

[1] Aqueous media, luminescence data is at 77 °K.

the luminescence data are for aqueous glucose-containing glasses at 77 °K.

As can be seen from Fig. 3, the fluorescence spectra of the aromatic amino acids are rather devoid of structure. The phosphorescence spectra all are composed of several partially resolved bands, however, and only the overall maxima are reported in Table 1. Tryptophan generally shows two very well resolved bands with peaks at roughly 405 and 435 nm. An additional peak is occasionally resolved in the vicinity of 455 nm but more often a broad plateau appears in this region.

The sum of the quantum yields for fluorescence (Φ_f) and for phosphorescence (Φ_p) is unity for tyrosine and phenylalanine at 77 °K, indicating that there are no appreciable non-radiative decay processes for these amino acids. Since $\Phi_f + \Phi_p \sim 0.89$ for tryptophan, some radiationless decay routes may be involved in this case.

The phosphorescence lifetimes are on the order of several seconds in each case which is consistent with a π,π^* phosphorescent triplet state. The highly forbidden nature of the phosphorescence transition ($S_0 \leftarrow T_0$) as indicated by the long phosphorescence lifetime makes direct optical excitation of the triplet states not feasible with anything less than a high power laser source.

The phosphorescence decay lifetimes of indole, tryptophan, and chymotrypsinogen have been measured [32] at about eight wavelengths in the range 400—500 nm. The phosphorescence lifetime was found to be independent of wavelength in each case. Since the polarization of the phosphorescence of indole and tryptophan also was found to be independent of wavelength [33], Volotovskii et al.[32] conclude that the entire phosphorescence spectrum results from a single electronic transition. In particular, it was concluded that the chymotrypsinogen phosphorescence contains only a contribution from the tryptophan triplet state.

The excitation spectra for both the fluorescence and phosphorescence of the aromatic amino acids coincide with the absorption spectra [27,13]. The same luminescence also can be induced by X-rays. Steen [34] has shown that aqueous solutions of tyrosine and tryptophan at 77 °K yield the same fluorescence and phosphorescence spectra upon excitation with UV light or X-rays. The phosphorescence lifetime is also unchanged but the phosphorescence to fluorescence ratio is greatly enhanced with X-irradiation. The apparent explanation is that the X-rays ionize the amino acid and the subsequent luminescence occurs upon a recombination process which results in a high probability of direct triplet production as well as that normally resulting from intersystem crossing from S_1. Thus, X-irradiation although generally less efficient than UV in producing luminescence, could provide a means of efficiently populating the phosphorescent triplet state in molecules where the quantum yield of intersystem crossing is too low for UV-induced luminescence to be observed.

2.1.1. Substituent Effects

The UV absorption spectra of tryptophan, tyrosine and phenylalanine are virtually identical to those of indole [35], phenol [36] and benzene [37] respectively. Close similarities also exist between the fluorescence and phosphorescence spectra of each pair, as well as between the phosphorescence lifetimes [8,28,38]. This and the electron spin resonance observations to be discussed in a later section indicate that the addition of a methyl group or other substituents of the type —CH_2R does not have a large effect on the measurable properties of the phosphorescent triplet state of the compounds being discussed here.

Weinryb and Steiner [39] examined the luminescence properties of a number of peptide derivatives of tryptophan and phenylalanine at room temperature and at 91 °K. No significant variations were found in the phosphorescence lifetimes of the tryptophan derivatives at 91 °K. Changes noted in the fluorescence quantum yields and lifetimes [39-41] upon substitution of side groups to the aromatic chromophores, ionization of amino acid chains and the presence or absence of peptide link ages are imperfectly understood.

2.1.2. Solvent Effects

An aromatic amino acid residue of a protein may be buried deep within the three dimensional structure of the protein, effectively shielded from the surrounding medium or it may lie on the surface highly exposed to the solvent medium. Furthermore, protein molecules *in vivo* are found in many different types of media ranging from the aqueous soup of cell cytoplasm to the highly ordered crystalline type environments of structural proteins. Thus, it is necessary to examine the luminescence properties of the aromatic amino acids in a wide variety of solvent media in order to begin to interpret protein emission characteristics.

Several authors have presented data on the quantum yields of fluorescence and phosphorescence of the aromatic amino acids and related compounds as functions of pH [26,31,42-44]. The quantum yields are found, in general, to be constant over the pH range of 3—8 but to vary quite dramatically at the extremes of pH. The relative phosphorescence intensities of tyrosine and tryptophan at 77 °K as functions of pH are reproduced in Fig. 5. The pH values where the phosphorescence intensities begin to vary are close to the pK's for ionization of the amino acids [45] (Table 2). However, in interpreting these results it should be

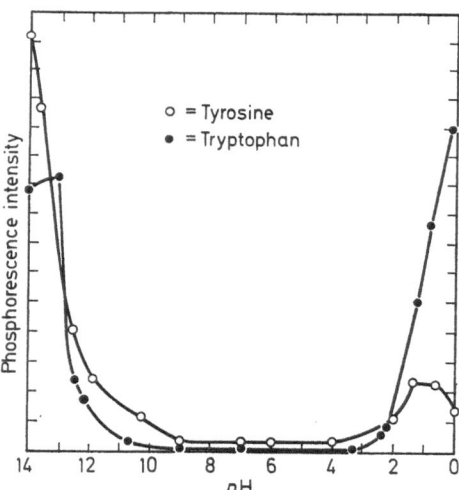

Fig. 5. Phosphorescence intensity of tryptophan and tyrosine as a function of pH. Concentration is 10^{-3} M in 0.5% glucose at 77 °K (from Truong, *et al.* [26]). [From Truong, Bersohn, Brumer, Luk, Tao: J. Biol. Chem. *242*, 2979 (1967), Fig.2.]

124

Table 2. pK values for ionization of the aromatic amino acids [38,45]

	pK$_1$(COOH)	pK$_2$(NH$_2$)	pK$_3$(OH)
Tryptophan	2.38	9.39	
Tyrosine	2.20	9.11	10.07
Phenylalanine	1.83	9.13	

remembered that the pK value in an excited state may be considerably different from that of the ground state of the molecule and also the pH of a solution measured at room temperature may not be a particularly meaningful quantity when the solution is frozen.

Augenstein and co-workers [31] have noted that the total emission yield of the aromatic amino acids at 77 °K is nearly independent of pH and only the ratio of phosphorescence to fluorescence changes. In order to account for the pH dependence of phosphorescence/fluorescence in tyrosine, where there is a large increase in this ratio above pH 9, Truong et al. [26] invoke an argument involving a shift in the energy level of a low-lying $^3n,\pi^*$ state relative to the lowest excited singlet state. The energy level schemes assumed for tyrosine and the tyrosinate ion are shown in Fig. 6. It subsequently follows that the enhanced phosphores-

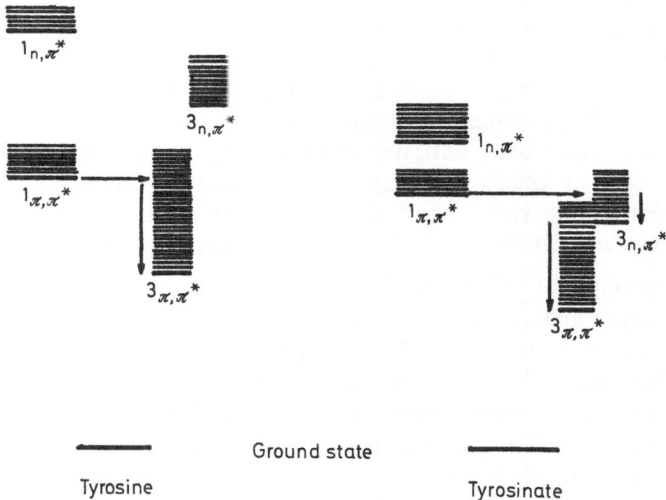

Fig. 6. Energy level scheme proposed for tyrosine and tyrosinate ion by Truong et al. [26]. Arrows represent radiationless intersystem crossing and internal conversion processes

125

cence intensity and the fluorescence quenching of the tyrosinate ion is caused by the more efficient intersystem crossing from $^1\pi,\pi^*$ to $^3n,\pi^*$ in tyrosinate ion as opposed to $^1\pi,\pi^*$ to $^3\pi,\pi^*$ in tyrosine [46]. The explanation of the quenching of tyrosine fluorescence by the carboxylate ion might involve the transfer of tyrosine proton to COO^- in the excited singlet state resulting in enhanced intersystem crossing to $^3n,\pi^*$. However, the mechanism must be more complex than the quenching of fluorescence by OH^-, since phosphorescence is quenched by COO^-, as well [26].

It has been observed also that at a given pH, an increase in ionic strength leads to an increase in the phosphorescence to fluorescence ratio in frozen solutions, again in such a way that the total emission yield is unchanged [31,43]. Since this "salt effect" was found to be similar for various singly charged electrolytes and added salt had no effect on the emission at room temperature, it was concluded [43] that the effect was brought about by changes in the solvent matrix surrounding the emissive molecules.

There are significant changes in both the wavelengths of emission and the fine structure of the luminescence of the aromatic amino acids and their parent molecules in going from a polar to a non-polar solvent [8,47]. Chignell and Gratzer [48] have carefully separated out the effects of hydrogen bonding and solvation upon addition of varying concentrations of polar solvent to p-cresol and indole in isooctane. The major change in the absorption spectra (and presumably in the emission spectra as well) is a red shift and loss of fine structure upon formation of a hydrogen-bonded chromophore, i.e. with the addition of just a small percentage of polar solvent. Upon further addition of polar solvent there is a continued but relatively minor shift due to the effect of solvent polarity changes on the already hydrogen bonded chromophore. The subsequent smaller solvent shift may be either a blue shift of a red shift depending upon the particular solvent mixture.

The significant implication of this work with respect to the luminescence of tryptophan and tyrosine in proteins is that although the aromatic chromophore may be buried in a hydrophobic region of the protein and well shielded from the solvent medium, it frequently is still involved in internal hydrogen bonding with neighboring amino acid side groups or with the polypeptide backbone. Hence, most of the vibrational structure of the absorption and emission is already masked by the spectral broadening caused by the hydrogen bonding and there would be only relatively minor differences in the emission of surface and buried aromatic chromophores due to differences in polarity and polarizability of the environments.

Konev [8] and co-workers have reported the luminescence properties of the aromatic amino acids and related compounds in a number of non-polar solvents and in solid films of polyvinyl alcohol. No surprising features are found in these media; the results in polyvinyl alcohol are very similar to those in frozen aqueous solutions whereas those in non-polar solvents incapable of forming hydrogen bonds differ by showing the expected blue shifts and additional fine structure.

The emission spectra from powders and "crystal suspensions" of the aromatic amino acids, however, are quite different from those shown in Figs. 3 and 4. The fluorescence and phosphorescence maxima of the powders are red shifted in each case by up to 50 nm and the phosphorescence lifetimes at 77 °K are found to be 1.5 sec for tryptophan, 0.4 sec for tyrosine and 0.5 sec for phenylalanine [30]. The phosphorescence lifetimes decrease markedly as the sample temperature increases [49]. In the powders the phosphorescence to fluorescence ratios are considerably smaller than in frozen aqueous solutions. The "crystal suspensions" at room temperature show even greater red shifts in the phosphorescence maxima, with no fine structure observed and with lifetimes on the order of 0.2 sec for each aromatic amino acid [50]. Although these results along with the phosphorescence excitation spectra bear little resemblence to those expected for the aromatic amino acids, Bogach *et al.* [50] point out the similarities to phosphorescence and excitation spectra of photoproducts formed in solutions of the aromatic amino acids at low temperatures [51,52].

2.1.3. Temperature Dependence

As we have mentioned previously, the phosphorescence of aromatic amino acids in solution is completely quenched at room temperature by rapid non-radiative processes. Non-radiative processes appear to be practically dormant at 77 °K since the sum of the fluorescence and phosphorescence quantum yields is close to unity for each aromatic amino acid (Table 1). As the temperature is raised the phosphorescence quantum yield begins to decrease drastically as solvent reorientation sets in. This occurs between 170° and 200 °K for frozen 0.5% glucose solutions [53]. A plot of the variation of the tryptophan phosphorescence quantum yield with temperature [54] is shown in Fig. 7.

2.2. Electron Paramagnetic Resonance

Detection of the phosphorescent triplet states in frozen solutions of aromatic molecules by electron paramagnetic resonance (EPR) was not accomplished until somewhat over a decade ago [55,56] shortly after

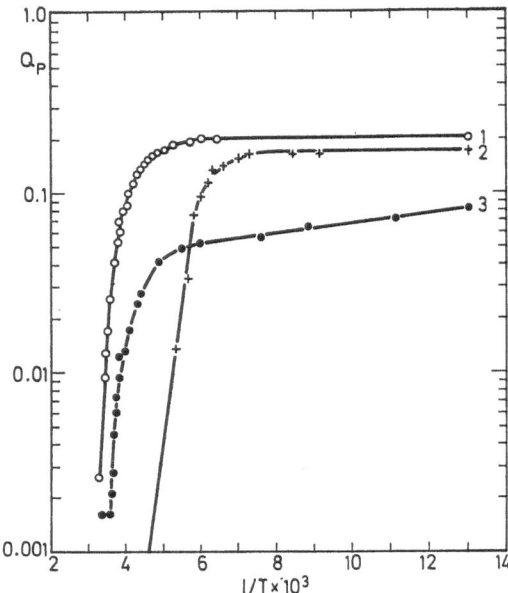

Fig. 7. The temperature-dependence of the phosphorescence quantum yields of tryptophan in (1) polyvinyl alcohol, (2) 0.5% glucose in water, and (3) trypsin (from Kuntz [54]).
[From Weinryb, Steiner: Excited states of proteins and nucleic acids, pp. 277–318. Fig. 3, p. 300. (1971).]

Hutchison and Mangum pioneered the technique with their studies of aromatic chromophores doped into single crystals [57,58]. The initial investigations [55,56] were limited to observations of $\Delta m_s = \pm 2$ "forbidden" transitions but the development of equipment with improved sensitivity rapidly led to the observation and interpretation of $\Delta m_s = \pm 1$ transitions of the triplet state in frozen solutions [59,60]. Observation of at least two $\Delta m_s = \pm 1$ transitions from frozen solutions allows the aromatic chromophore to be characterized by the parameters $|D|$ and $|E|$ which specify the zero-field splitting (ZFS) between the triplet sublevels [61,62]. The $\Delta m_s = \pm 2$ transition by itself will yield only a root square (rms) zero-field parameter, $D' = \sqrt{D^2 + 3E^2}$.

Among the first molecules for which triplet state EPR signals were detected in randomly oriented frozen solutions were indole [63] and tryptophan [64]. The aromatic amino acids are well suited to triplet EPR studies by virtue of their long-lived phosphorescent states which allow large steady-state triplet populations to be achieved with continuous optical pumping. The triplet states of the aromatic amino acids and related compounds were initially characterized by their rms zero-field parameters D' and the decay times of the EPR signals after switching off the exciting light [65-67]. The EPR decay times were found to be comparable with the phosphorescence decay times [66].).

In later studies observations of the $\Delta m_s = \pm 1$ EPR signals of the aromatic amino acids also were reported [68-70]. The zero-field parameters are collected in Table 3 along with those of triplet states of

Table 3. Triplet state zero-field splitting parameters[1]) of the aromatic amino acids and related compounds [70]

	D/hc (cm^{-1})	E/hc (cm^{-1})	D'/hc (cm^{-1})
Indole	0.1011	0.0416	0.1241
Skatole	0.0965	0.0436	0.1225
Tryptophan	0.0984	0.0410	0.1213
Phenol	0.1352	0.0451	0.1561
p-Cresol	0.1251	0.0590	0.1615
Tyrosine	0.1301	0.0558	0.1621
Phenolate ion	0.1133	0.0337	0.1275
p-Cresolate ion	0.1071	0.0540	0.1422
Tyrosinate ion	0.1080	0.0540	0.1429
Benzene	0.1581	0.0044	0.1583
Toluene	0.1454	0.0250	0.1517
Phenylalanine	0.1475	0.0439	0.1659

[1]) The D and E parameters are those appearing in the zero-field spin hamiltonian $D(S_z^2 - \frac{2}{3}) + E(S_x^2 - S_y^2)$, where z is the out-of-plane principal axis. D is assumed positive, and the relative signs of D and E are obtained from calculations of the zero-field splitting [85]. The axis system for indole-related triplets is shown in Fig. 8A, while that for the phenol derivatives has the principal x-axis coincident with the short axis of the molecule. The principal axis perpendicular to the molecular plane was determined by magnetophotoselection [70].

related molecules. It is seen that addition of a substituent of the type —CH$_2$R has little effect on the zero-field parameters except in the case of benzene where addition of a substituent removes the hexagonal symmetry axis. The E parameter is equal to zero if the molecule has C$_3$ or higher symmetry [46]. The small but non-zero value of E observed for benzene is attributed to a Jahn-Teller distortion in the triplet state [71].

The small changes in ZFS in the other cases are in accord with the observations of Smaller et al. [72] of correlations between the splittings and π-charge redistributions induced by the substituents [73].

Along with the optically excited triplet states, EPR also results in the detection of free radicals formed upon irradiation of frozen solutions of the aromatic amino acids [74,75]. In all aqueous and alcoholic media examined solvent radicals were induced by the presence of an aromatic solute. Once formed the radicals are stable after extinction of the exciting light until the solution is warmed to temperatures at which solvent reorientation begins to take place [74].

Azizova et al. [76] showed that the production of free radicals involved a mechanism in which the triplet level of the aromatic chromophore appears as an intermediate state. The mechanism proposed electron photoejection after absorption of a second photon by an aromatic molecule already excited to its phosphorescent triplet state [77,78]. In acidic media the photoejected electron can be trapped by a proton and the EPR signal of hydrogen atoms is observed. In basic media the EPR signal from $O(-)$ ions is observed [79]. In some instances secondary excitation with visible light induces radical recombination by freeing the trapped electrons [79].

The major interest in such radical EPR studies lies in identifying the role of the triplet states in photobiological processes. The establishment of triplet state intermediates has now been confirmed in a wide variety of photoionization and sensitization reactions [76,80,81] in frozen solutions of nucleic acid bases, porphyrins including chlorophyll and, of course, the aromatic amino acids [82]. Comparisons of the photo-induced products formed in these systems with observations in vivo provides strong evidence that the triplet state is an essential intermediate in many photobiological processes.

The specification of the polarizations of the electronic transitions in the aromatic amino acids is of interest in describing the excited states of these molecules and in considerations of energy transfer between the aromatic residues of peptides and proteins. The technique of magneto-photoselection [83,84] (measuring the relative intensities of the $\Delta m_s = \pm 1$ transitions as a function of the orientation of the electric vector of linearly polarized exciting light relative to the external magnetic field) has been used to find the relative orientations between the electronic transition moments and the principal magnetic axes of tyrosine and tryptophan [70]. These results were then used in conjunction with calculations of the principal magnetic axes to deduce the absolute molecular orientations of the electronic transitions [70,85]. The calculated [70,85] principal magnetic zero-field axes system of tryptophan is given in Fig. 8.

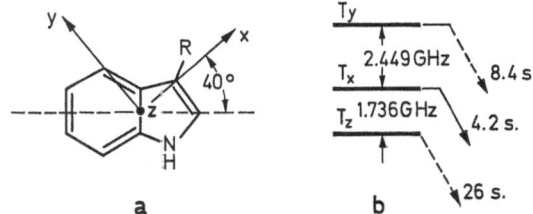

Fig. 8. (a) The principal axes system of the triplet state of tryptophan according to the calculations of Zuclich [70,85]. The x- and y-axes of Fig. 5(b) of Ref. [101] are incorrectly labeled, and should be interchanged.
(b) Zero-field level ordering of tryptophan triplet from magneto-photoselection measurements [70,85]. Zero-field frequencies, decay constants, and radiative pattern are from ODMR measurements [99,101]. Radiationless decay is represented by dashed arrows

2.3. Optical Detection of Magnetic Resonance

With the development in the past several years of the technique of optically detected magnetic resonance (ODMR) [86-88], a powerful new technique has become available for studying the zero-field splittings as well as the kinetic parameters associated with populating and depopulating of the phosphorescent triplet state [89-93]. ODMR (also referred to in the literature as phosphorescence microwave double resonance, or PMDR) [94] can be used to determine the total decay rate constants, k_i, of each triplet sublevel, T_i, the spin-lattice relaxation (SLR) rate constants between sublevels, W_{ij}, as well as to provide relative values for the populating rates, P_i, radiative rate constants, k_i^r, and steady-state populations, N_i^0. These rate parameters are illustrated in Fig. 9.

Although the initial measurements of ODMR [86-88] were made in the conventional manner, (fixed microwave frequency source and variable externally-applied Zeeman field) nearly all measurements on the aromatic amino acid triplet states have been made at zero applied field using a variable frequency microwave source. Consequently, we will focus on the zero-field experimental method in this section.

The essential features of ODMR are based upon the fact that, in general, the magnetic sublevels of the triplet state which we label T_x, T_y, and T_z have distinct and distinguishable properties. Selective intra-molecular spin-orbit coupling routes cause the sublevels to mix to a differing degree with excited singlet states. Since intersystem crossing (radiative and radiationless) is extremely sensitive to the singlet character of the triplet sublevel, selective populating and decay routes are established [95]. At temperatures at which most phosphorescence decay

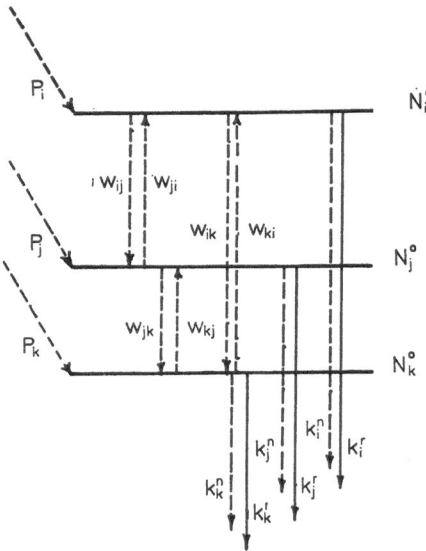

Fig. 9. Energy level diagram of a phosphorescent triplet state in zero field illustrating the dynamic processes which determine the steady-state populations (N_i^0) of the magnetic sublevels (T_i). The symbols are defined in the text. Broken arrows represent radiationless processes while normal arrows represent radiative ones. $k_i = k_i^n + k_i^r$

experiments are carried out (77 °K), the distinct properties of the triplet sublevels are not noticeable since rapid spin-lattice relaxation (phonon-induced transitions among the magnetic sublevels) maintains the sublevel populations in a Boltzmann equilibrium. The sublevels are thus nearly equally populated at all times and the triplet state decays as would a single excited state with a decay constant given by the average,

$$\bar{\varrho} = \frac{1}{3}\,(k_x + k_y + k_z) \tag{1}$$

where the k_i are the individual decay constants of the sublevels. The distinct properties of the triplet sublevels are revealed when the temperature is reduced sufficiently that the spin-lattice relaxation rates become slower than the decay rates of the sublevels. At pumped helium temperatures, ca. 1.1—1.2 °K, spin-lattice relaxation frequently becomes negligible relative to the decay rates, and the sublevels are then effectively decoupled from one another. Under these conditions, the high-temperature simple exponential decay of the triplet state becomes complex — a superposition of three individual exponential decays. If each of the sublevels is radiative, the phosphorescence will be observed to decay as the sum of three first order decays, with decay constants k_x, k_y, and k_z. In general, the populations (N_i) of the sublevels are given by the solutions to the coupled differential equations

$$dN_i/dt = P_i - (k_i + W_{ij} + W_{ik}) N_i + W_{ji}N_j + W_{ki}N_k \qquad (2)$$

where P_i is the populating rate, and W_{ij} is the spin-lattice relaxation rate constant governing the transitions $T_i \rightarrow T_j$. In the absence of significant spin-lattice relaxation, (2) simplifies to

$$dN_i/dt = P_i - k_iN_i . \qquad (3)$$

In the steady state of constant optical pumping

$$N_i^0 = P_i/k_i \qquad (4)$$

where N_i^0 is the steady-state population of T_i. In general, the ratios P_i/k_i and therefore the N_i^0 will be different for $i = x,y,z$, a condition referred to as spin alignment. The steady-state phosphorescence intensity is then a sum over the sublevels,

$$I_0 = S\sum_i k_i^r N_i^0 , \qquad (5)$$

since we are not able generally to optically resolve the emissions from the individual sublevels; the zero-field splitting is usually ca. 0.1 cm^{-1}. In (5) k_i^r is the radiative rate constant of T_i while S is a constant depending upon the apparatus. If at an arbitrary time, $t = 0$, we suddenly saturate (equalize) the populations of T_i and T_j with a radiofrequency magnetic field at the resonance frequency the observed change in the phosphorescence intensity at $t = 0$ will be

$$\Delta I (0) = \frac{1}{2} S(N_i^0 - N_j^0) (k_j^r - k_i^r) . \qquad (6)$$

Eq. (6) summarizes the conditions required for ODMR, i.e., that there be spin alignment of the triplet state ($N_i^0 \neq N_j^0$), and that the radiative rate constants of the sublevels undergoing resonance must differ.

2.3.1. Fast Passage ODMR

It is easily shown [93] that if the levels are saturated with a resonance microwave pulse or fast-passage of negligible duration compared with the lifetime of either sublevel, the equation governing the time dependence of ΔI is

$$\Delta I(t) = \frac{1}{2}S(N_i^0 - N_j^0)(k_j^r\, e^{-k_j t} - k_i^r\, e^{-k_i t}) = A e^{-k_j t} - B e^{-k_i t}. \quad (7)$$

The response of the phosphorescence to a short saturating microwave pulse during constant optical pumping is a discontinuous jump of amplitude A—B followed by a decay back to the steady state which is a linear combination of two exponentials with lifetimes equal to the decay constants of the saturated levels. An experimental observation of this transient is shown in Fig. 10 for the triplet state of tryptophan

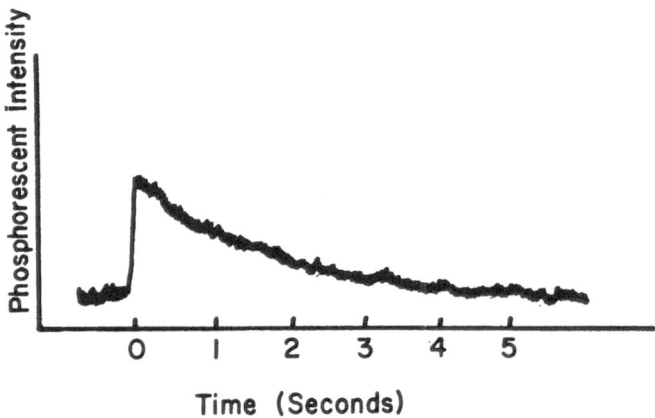

Time (Seconds)

Fig. 10. Transient response of the phosphorescence intensity of bovine serum albumin monitored at 414 nm to a microwave fast-passage magnetic resonance transition which occurs at $t = 0$. The magnetic resonance, which is centered at 1.657 GHz is due to the tryptophan $T_z \rightleftarrows T_x$ transition (see Fig. 8). The temperature is 1.3 °K, no external magnetic field is present, the solvent is 50% ethylene glycol-water, and the sample is continuously optically pumped. The transient decays as a single exponential since only T_x is radiative. If both T_x and T_z had been radiative, a response such as shown in Fig. 1 of Winscom and Maki [93] would have been observed. (From Zuclich et al. [100b])

in bovine serum albumin (BSA) for which only one of the saturated levels is radiative. It should be pointed out that analysis of the transient for a case where both levels are radiative gives the relative radiative rate constants as the ratio $A/B = k_j^r/k_i^r$. Analysis of these transient responses to saturation for at least two of the three possible zero-field transitions yield all three decay constants, k_i, the relative radiative rate constants, k_i^r, the relative steady state populations, N_i^0, and the relative populating rates, P_i. [93]

Several other ODMR approaches have been described [92,96–98] which yield the same detailed kinetic information (also in the absence of spin-lattice relaxation), but a complete description of these methods is beyond the scope of this review.

2.3.2. Effects of Spin-Lattice Relaxation

In considering the ODMR results of such long-lived phosphorescent species as the aromatic amino acids it is likely that spin-lattice relaxation is not negligible even at the lowest temperatures achieved in any ODMR experiments to date (1.1–1.2 °K). The fast-passage ODMR method and alternative techniques mentioned in the preceding section all have assumed that spin-lattice relaxation between the triplet sublevels could be neglected. Very recently, however, Zuclich et al. [99] have presented a method for the determination of spin-lattice relaxation rates and for calculating their effects on ODMR responses.

The technique rests on monitoring the phosphorescence decay (after the exciting light is turned off) while continuously saturating one or more of the triplet sublevel transitions. Upon extinguishing the exciting source, Eq. (2) can be rewritten in the form

$$dN_i(t)/dt = - K_i N_i(t) + W_k N_j(t) + W_j N_k(t) \tag{8}$$

where $K_i = k_i + W_k + W_j$ and $W_{ij} = W_{ji} \equiv W_k$.

The assumption that $W_{ij} = W_{ji}$ is not strictly valid. In a two-level system, they are related by the Boltzmann factor. This is also true in a multilevel system provided that relaxation occurs by direct absorption and emission of resonant phonons in thermal equilibrium with the lattice. The Boltzmann factor is nearly unity for the conditions of the ODMR measurements on the aromatic amino acids. Thus, we assume that no large errors are introduced by the assumption that $W_{ij} = W_{ji}$. A more detailed discussion of the effects of this approximation is given in Ref. [99].

The general solution for the set of three coupled differential equations represented by Eq. (8) is given by

$$N_i(t) = \sum_{n=1}^{3} c_{in}\, e^{-\varrho_n t} \tag{9}$$

where the decay constants, ϱ_n, are the solutions of a cubic polynomial equation, and consequently rather unwieldy functions of the k's and W's. It is apparent that in the presence of significant spin-lattice relaxation, *each* sublevel decays as the sum of three first-order processes with identical decay constants. The phosphorescence decay is thus the sum of three exponential terms.

$$I(t) = S \sum_{i,n=1}^{3} k_i^r\, c_{in}\, e^{-\varrho_n t} \tag{10}$$

In the case where the ith and jth sublevels are continuously saturated during the phosphorescence decay, the rate equations are

$$dN(t)/dt = -\frac{1}{2}(k_i + k_j + W_j + W_i)\, N(t) + \frac{1}{2}(W_j + W_i)N_k(t) \tag{11a}$$

$$dN_k(t)/dt = -(k_k + W_j + W_i)N_k(t) + (W_j + W_i)N(t) \tag{11b}$$

where $N_i(t) = N_j(t) \equiv N(t)$. The problem is now reduced to a system of two coupled differential equations with a general solution for the decay constants given by the roots of a quadratic equation. Eq. (10) correspondingly simplifies to

$$I(t) = S \sum_{m,n=1}^{2} k_m^r\, c_{mn}\, e^{-\varrho_n t} \tag{12}$$

where the radiative rate constants are identified as $(k_i^r + k_j^r)$ and k_k^r. The phosphorescence decay is the sum of two exponentials. The sum, ϱ_s, and product, ϱ_p, of the phosphorescence decay constants are readily shown [99] to be given by the expressions

$$\varrho_s^k = \frac{1}{2}(k_i + k_j) + k_k + \frac{3}{2}(W_i + W_j) \tag{13}$$

$$\varrho_{\mathrm{p}}^{k} = \frac{1}{2}k_{k}(k_{i} + k_{j}) + \frac{1}{2}(k_{i} + k_{j} + k_{k})(W_{i} + W_{j}) \tag{14}$$

where the superscript, k, on ϱ_{s} and ϱ_{p} indicates microwave saturation between the ith and jth sublevels. There are three such pairs of equations, each corresponding to microwave saturation between one of the three sublevel pairs. However, an additional independent equation is required in order to solve for absolute values of the k_{i} and W_{i}. This is provided by monitoring the phosphorescence decay in the presence of simultaneous microwave saturation between two pairs of sublevels. The decay is then a single exponential with a rate constant, $\bar{\varrho}$, given by Eq. (1).

One may not generally assume (*vide infra*) that there is no temperature dependence of the total decay rates so $\bar{\varrho}$ should not be obtained from the high temperature (*i.e.* 77 °K) phosphorescence decay rate.

Eqs. (1), (13) and (14) can be used to solve for the k_{i} and W_{i} in terms of the experimentally obtainable quantities $\bar{\varrho}$, ϱ_{s} and ϱ_{p}. Thus,

$$k_{k} = \bar{\varrho} \pm \sqrt{2\left(\bar{\varrho}\,\varrho_{\mathrm{s}}^{\mathrm{k}} - \bar{\varrho}^{2} - \varrho_{\mathrm{p}}^{k}\right)}, \tag{15}$$

$$W_{i} + W_{j} = \frac{2}{3}\varrho_{\mathrm{s}}^{k} - \bar{\varrho} - \frac{1}{3}k_{k}. \tag{16}$$

Once the k_{i} and W_{i} are found from the six equations represented by (15) and (16), the relative k_{i}^{r}, P_{i}, and N_{i}^{0} are then easily obtained by simple extension of the equations developed for negligible spin-lattice relaxation.

One important feature which emerges from the theory described above is that the decay constants observed in the phosphorescence decay measurements should correspond exactly to the fast-passage ODMR decay constants (Section 2.3.1.) even in the presence of spin-lattice relaxation.

2.3.3. ODMR of Indole and Tryptophan

To our knowledge the first application of ODMR to a molecule of biological importance was made by Kwiram [100a] who in 1970 reported the optical detection of magnetic resonance signals from the tryptophan moiety of lysozyme. A more recent report dealt with frozen glassy solutions of tryptophan, tyrosine, and the protein, bovine serum albumin (BSA) [100b]. See Fig. 10. Discrepancies between the lifetimes involved in the phosphorescence decay and the fast-passage ODMR responses

A. H. Maki and J. A. Zuclich

of these molecules [101] suggested that non-negligible spin-lattice relaxation processes were involved in the triplet state observations even at 1.25 °K. It is now clear [99] that not only are spin-lattice relaxation processes non-negligible but also that they are different in the fast-passage ODMR measurements (which are done during continuous optical pumping) and in the phosphorescence decay measurements (where optical pumping is interrupted). Inefficient dissipation of the heat resulting from optical pumping to the helium bath in the former experiments results in a higher lattice temperature and larger spin-lattice relaxation rates. Thus, the observed fast-passage ODMR decays are invariably faster than the phosphorescence decays [99,101]. The method described in Section 2.3.2. was applied to indole and tryptophan at 1.25 °K with the results for the total sublevel decay constants, and spin-lattice relaxation rate constants given in Table 4. Only the T_x sublevel of tryptophan is found to be radiative [100b,101]; the others decay by radiationless processes, including spin-lattice relaxation via the radiative T_x sublevel. Intersystem crossing from S_1 is mainly to the T_x and T_y sublevels [99].

Table 4. Sublevel decay constants and spin-lattice relaxation rates[1]

Sample	Total sublevel decay rates			Spin-lattice relaxation rates		
	k_x (sec^{-1})	k_y (sec^{-1})	k_z (sec^{-1})	W_z (sec^{-1})	W_y (sec^{-1})	W_x (sec^{-1})
Indole (single crystal) excitation, 285 nm detection, 404.3 nm	0.370	0.167	0.079	0.003	−0.001	0.029
Indole (indan) excitation, 279 nm detection, 401.6 nm	0.353	0.158	0.067	0.014	0.032	0.010
Tryptophan (ethylene glycol—H$_2$O) excitation, 295 nm detection, 406.0 nm	0.240	0.119	0.038	−0.002	0.036	0.036

[1] Method of Ref.[99], $T=1.25$°K. Spin-lattice relaxation rates are in absence of optical pumping. Small negative W's are zero within estimated accuracy of ±0.005 sec^{-1}.

At 77 °K, where spin-lattice relaxation maintains the triplet sublevel populations essentially the same, the fraction of the tryptophan triplet population which decays radiatively is

$$F_R = (1 - f_x)k_x/(k_x + k_y + k_z) \qquad (17)$$

138

which from Table 4 is $F_R = 0.605\,(1-f_x)$. In Eq.(17), f_x is the probability that the T_x sublevel decays radiationlessly. The quantum yield for phosphorescence is consequently

$$\Phi_p = 0.605\,(1-f_x)\,\Phi_{isc} \qquad (18)$$

where Φ_{isc} is the quantum yield for intersystem crossing from S_1. Φ_{isc} is also given by

$$\Phi_{isc} = 1 - \Phi_f - \Phi_d = 0.28 - \Phi_d \qquad (19)$$

where Φ_d is the quantum yield for radiationless internal conversion from the fluorescent state, and $\Phi_f = 0.72$ is obtained from Table 1. Substitution of $\Phi_p = 0.17$ (Table 1) into Eq. 18, and combination with Eq. 19 gives

$$(0.28 - \Phi_d)\,(1 - f_x) = 0.28 \qquad (20)$$

Eq. 20 is only valid if $\Phi_d \sim 0$, and $f_x \sim 0$. This implies that radiationless processes occur from the triplet state only, and furthermore that the radiationless processes occur only from the T_y and T_z sublevels. The T_x sublevel decays predominantly by radiative processes at 1.2 °K. This little calculation assumes, of course, that the sublevel decay constants measured at 1.2 °K may be applied to quantum yield data at 77 °K. As we will discuss in more detail in a later section, the low temperature decay constants predict a somewhat longer triplet lifetime than is observed experimentally at 77 °K which indicates the presence of thermally-activated radiationless quenching of the tryptophan triplet even at liquid nitrogen temperature. This effect is rather small, however, and should not affect the general conclusions reached above concerning the energy degradation pattern of the tryptophan excited singlet state. The decay pattern of the triplet sublevels of tryptophan is shown in Fig. 8.

In order to more fully characterize the properties of the triplet states of tryptophan residues at different sites in proteins, detailed ODMR studies of indole and tryptophan in various environments were carried out [99,101,102]. It was found that the triplet energy, as measured by the origin of the phosphorescence spectrum, as well as the zero-field splitting, of these molecules are very sensitive to environment. The phosphorescence emissions from crystalline indole and of indole and tryptophan in frozen glasses are broad ($\sim$150—300 cm^{-1} half-width for the 0—0 bands) at liquid He temperatures and originate from an inhomogeneous distribution of traps. Indole in indan, however, exhibits a

Shpol'skii effect [103] showing a number of optically resolvable origins (different trap sites) with ~12 cm^{-1} halfwidths for the 0—0 bands. ODMR and optical electron-electron double resonance [104] experiments showed that some of the optical origins themselves originate from multiple sites.

ODMR linewidths of indole in single crystals and frozen solutions are on the order of 100 MHz. Hole burning experiments [105] were used [102] to show that the ODMR lines have an apparent homogeneous linewidth of ~10 MHz. This width is comparable to that expected from "forbidden" satellite transitions caused by the ^{14}N quadrupole interaction.

Results for indole —h_1, indole —d_1 and N-methyl indole are similar indicating that the sensitivity of the indole triplet state properties to the environment does not involve direct interactions of the N—H bond [102].

3. Polypeptides

The incorporation of an aromatic amino acid into a protein may have considerable influence on the emissive properties of the chromophore and may result in electronic energy transfer between the various aromatic residues of the protein. However, since the protein itself is generally a rather large and complicated molecule (molecular weights range from the order of 10^4 to 10^6) and the three dimensional structure is frequently unknown, the analysis of effects involving the aromatic chromophores is an exceedingly difficult problem. Fortunately, a large number of small peptides of known composition and structure are available as model systems. This includes synthetic and naturally occurring peptides (e.g. many hormones and antibiotics) which contain only one of the aromatic amino acids or any combination of two aromatic chromophores. Also available for study are higher molecular weight homo and hetero polypeptides of the aromatic amino acids.

The incorporation of tryptophan into a polypeptide does not dramatically alter its luminescence properties [28b]. The shape of the fluorescence spectra remains the same but there are shifts reflecting the degree of exposure of the tryptophan to the solvent medium [106,107]. A red shift of the fluorescence maximum of tryptophan to ~350 nm results from denaturation of a polypeptide, or protein, at room temperature. The shift depends upon the ability of solvent molecules to reorient during the singlet lifetime, however, and thus disappears in rigid media. The fluorescence [107,108] and phosphorescence [39] quantum yields are somewhat reduced from those of free tryptophan and depend to a small extent on the nature of the surrounding amino acid residues, particularly

140

the presence of histidine and lysine. There are minor changes in the quantum yields and Stokes' shifts with change in polypeptide conformation [109] due to changes in the immediate environment (including degree of exposure to solvent) of the emitting residues.

Tyrosine, like tryptophan, does not exhibit major changes in its luminescence spectrum when incorporated into peptides. Unlike tryptophan, however, there are large reductions in the fluorescence [110] and phosphorescence [111] quantum yields. It has been concluded from model system studies that $COO^{(-)}$ and NH_2 groups contribute to tyrosine quenching in polypeptides [112]. On the other hand, $COOH$ [113] and NH_3^+ groups [28b] may quench tryptophan luminescence but tryptophan is apparently less sensitive to neighboring groups in the polypeptide chain.

It has been noted also that tyrosine luminescence is strongly affected by the presence of an adjacent disulfide linkage [111]. The tyrosine phosphorescence shows a red shift and a greatly reduced quantum yield when adjacent to a disulfide group [111,114]. Using thioctic acid as an acceptor, it has been shown recently [115] that direct electron transfer occurs from the tyrosine triplet state to the disulfide linkage resulting in the formation of $RSSR^{(-)}$ ions. $RSSR^{(-)}$ ions are formed by trapping of e^- (aq.) ejected from the triplet state of tyrosine as well. Photodeactivation of enzymes such as ribonuclease A may well involve electron transfer to S—S linkages from tyrosine triplets followed by disruption of R_1S—$SR_2^{(-)}$ according to R_1S—$SR_2^{(-)} \rightarrow R_1S. + SR_2^{(-)}$.

Several bacterial antibiotics which contain phenylalanine as their only aromatic chromophore have been shown to exhibit luminescence spectra virtually identical to that of free phenylalanine [28b] although in one case [106] more highly resolved fine structure was attributed to the phenylalanine being in a hydrophobic environment. Phosphorescence lifetimes and fluorescence to phosphorescence ratios show only minor variations for a variety of phenylalanine-containing peptides [39].

The total emission and phosphorescence spectra of homopolymers of the three aromatic amino acids are reproduced in Fig. 11. The spectra are recorded at 77 °K using diglyme as a solvent [116]. The luminescence properties are summarized in Table 5. Comparisons can be made with the data for the free aromatic amino acids in Table 1 in analyzing the effects of incorporating the amino acids into the homopolymers which have well established regular three dimensional conformations [117,118].

The shapes of the absorption, fluorescence and phosphorescence spectra of the homopolymers are basically unchanged from those of the monomers and for the most part only minor shifts are noted in the wavelengths. The molar extinction coefficients of the polymers indicate a *hypo*chromic effect in poly-L-tryptophan [119] and poly-L-tyrosine [120] but a *hyper*chromic effect in poly-L-phenylalanine [121].

141

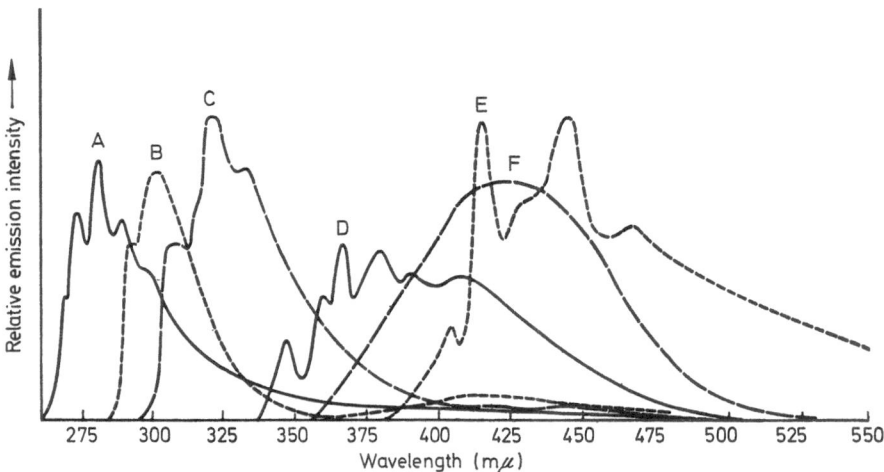

Fig. 11. Total emission and phosphorescence spectra of aromatic polyamino acids in diglyme solvent at 77 °K. Amino acid concentration is 5 × 10⁻⁴ M. Total emission spectra: (A) poly-L-phenylalanine (10 mv); (B) poly-L-tyrosine (100 mv); (C) poly-L-tryptophan (200 mv). Phosphorescence spectra: (D) poly-L-phenylalanine (1 mv); (E) poly-L-tryptophan (1 mv) (F) poly-L-tyrosine (10 mv). The amplifier gain setting is given in parentheses. (From Longworth [116]).
[From Longworth: Excited states of proteins and nucleic acids, pp. 319—484. Fig. 10B, p. 338 (1971).]

Table 5. Absorption and luminescence data for homoaromatic polypeptides [116]

	Poly-L-tryptophan	Poly-L-tyrosine	Poly-L-phenylalanine
Absorption maximum (nm)	282	279	259
Molar extinction coefficient at maximum	568	145	223
Fluorescence maximum (nm)	320	289	282
Fluorescence quantum yield	0.70	0.35	0.05
Phosphorescence maximum (nm)	445	415	367
Phosphorescence quantum yield	0.01	0.15	0.004
Phosphorescence lifetime (sec)	2.85	2.80	—

The quantum yields of phosphorescence of the homopolymers are drastically reduced relative to the monomers. Since the phosphorescence lifetimes and fluorescence yields are not as severely affected in poly-L-tryptophan and poly-L-tyrosine, incorporation into the regular poly-

142

peptide chains might inhibit intersystem crossing without affecting the internal conversion rates. More likely, the reduction of the phosphorescence quantum yield results from the rapid quenching of mobile triplet excitations (*i.e.* by paramagnetic impurities, and/or by triplet-triplet annihilation). The small yield of triplets with normal decay lifetimes could originate from immobilized trap sites in the otherwise regular polypeptide structure.

Dimers of tyrosine are readily formed as a photoproduct upon irradiation of poly-L-tyrosine [122]. Possibly the mechanism involves the initial formation of the tyrosine triplet state with subsequent photoejection of an electron, and the loss of a proton to form a neutral phenoxyl radical which attacks a neighboring tyrosine to form the 0—0'-dihydroxy diphenyl compound [28b].

3.1. Energy Transfer in Polypeptides

Small peptides and aromatic amino acid copolymers have proved to be excellent model systems for elucidating pathways of energy transfer between the aromatic residues of proteins.

The Förster mechanism for non-trivial transfer of excitation energy is well established [123]. Transfer may occur over distances as large as 100 Å or more, provided there is sufficient overlap between the emission band of the donor and absorption band of the acceptor. Weber showed by means of fluorescence polarization studies that transfer between phenol molecules and between indole molecules as well as phenol → indole transfer occurs over distances of $\sim$16 Å [124,125]. EPR studies of frozen solutions indicated that energy transfer occurs between tyrosine and tryptophan over distances of this order or larger [66,126].

In view of the occurrence of energy transfer over large distances it is not surprising that transfer is extremely efficient in small peptides and that the luminescence and magnetic resonance properties in such cases are largely those associated with just one of the aromatic amino acids. In the case of L-tryptophyl-L-tyrosine (trp-tyr), the fluorescence [111, 127], phosphorescence [127,128], and EPR [128] spectra are due solely to trytophan. Cassen and Kearns have used the excitation difference spectra technique [129,130] to show that tyr → trp energy transfer occurs with essentially unit efficiency at the singlet level [131].

For other small peptides the transfer is in the direction indicated by the ordering of the fluorescent singlet and the phosphorescent triplet energy levels. The ordering of both the singlet and the triplet levels is:

$$\text{phenylalanine} > \text{tyrosine} > \text{tryptophan.}$$

However, the tyrosine negative ion has a fluorescent singlet level lower than that of tryptophan but a phosphorescent triplet level lying between those of tyrosine and tryptophan. Thus, in trp—tyr$^{(-)}$, at the singlet level trp → tyr$^{(-)}$ transfer occurs while at the triplet level the converse is true [127,132]. Edelhoch *et al.*[133] monitored the trp → tyr$^{(-)}$ singlet-singlet transfer for the series of compounds trp—(gly)$_n$—tyr$^{(-)}$ where $n = 0—4$ (gly, glycine is an aliphatic amino acid). They found the degree of quenching of tryptophan fluorescence fell from 85% for the dipeptide to 50% for trp—(gly)$_4$—tyr$^{(-)}$.

The ionization of tyrosine provides an interesting case for energy transfer studies in homopolymers of tyrosine. Considerable effort has been expended in monitoring the tyr → tyr$^{(-)}$ transfer as a function of the percentage of ionization in (tyr)$_n$. The efficiency of transfer was found to be 100% in (tyr)$_2$ and (tyr)$_3$ when an average of one tyrosine is ionized [134]. In (tyr)$_6$ the transfer is less than 100% efficient with one ionized tyrosine, and the tyrosine fluorescence and phosphorescence are quenched to the same extent [127]. The same holds true for partially ionized poly-L-tyrosine [135,136] so that the occurrence of triplet-triplet energy transfer appeared doubtful. The possibility of triplet migration along the polypeptide chain was not ruled out, however, because of the occurrence of intramolecular migration of the singlet state over ∼100 residues in poly-L-tyrosine [135-139]. Thus, triplet state energy would have had to exhibit a similar range of migration before triplet-triplet transfer could have been detected. The occurrence of triplet-triplet energy transfer in unionized poly-L-tyrosine was eventually established by the detection of delayed fluorescence resulting from triplet-triplet annihilation [140,141]. This delayed fluorescence can account at least partially for the reduced Φ_p in poly-L-tyrosine. Tyr → tyr$^{(-)}$ triplet transfer has not been observed, however.

Singlet migration in poly-L-tryptophan was observed to occur to a similar extent as in poly-L-tyrosine but no delayed fluorescence was found, thus failing to confirm triplet migration [28b]. Fluorescence decay measurements on a *short* time scale (rather than a time scale comparable with typical rotating sector speeds) could conceivably reveal a *rapid* delayed fluorescence caused by the annihilation of short-lived mobile triplet excitations. Delayed fluorescence with a relatively long lifetime but complex decay kinetics (τ ranging between 89 msec and 2 sec) has been observed from poly-DL-tryptophan [28b]. This long-lived fluorescence component may be the result of relatively low triplet mobility in what is probably a random copolymer of the D- and L-stereoisomers. The regular structure of a homopolymer would be more consistent with high triplet mobility and possibly a concomitant short delayed fluorescence lifetime. In copolymers of tyrosine and tryptophan the tyr → trp transfer is found to be complete as might be expected from the large range of the singlet energy migration in poly-L-tyrosine.

4. Proteins

As has been stated previously, the absorptive and emissive properties of proteins are basically those of the aromatic amino acids. There are some changes in these properties due solely to the incorporation of the aromatic amino acids into peptide chains and these have been discussed in the preceding section. Even taking these changes into account, however, the luminescence properties of most proteins cannot be completely described as composites of the contributions from each of their aromatic residues. Among proteins with similar aromatic amino acid content there may be differences in the shape and wavelengths of the luminescence, the fluorescence and phosphorescence quantum yields and lifetimes, as well as differing effects of varying the solvent medium. Thus, the behavior of each aromatic residue depends somewhat on the exact three dimensional environment conferred upon it by the overall protein conformation and therefore is an innate probe whose properties describe the protein site where it sits and which can be used to monitor changes at that site.

4.1. Phosphorescence

A terminology for luminescence classification of proteins which specifies the presence of tryptophan (Class B proteins) or the absence of tryptophan but the presence of tyrosine (Class A proteins) is commonly used in the literature [106]. Since the presence of one or more tryptophan residues frequently dictates an overwhelming tryptophan dominance of both the fluorescence and phosphorescence this classification is also a statement of whether the protein luminescence is basically that of tryptophan or of tyrosine. A significant component of tyrosine emission is often found in Class B proteins [142-146] and in at least one case appears to dominate the fluorescence [28b]. Collagen, a tryptophan-free structural protein with a 3:1 phenylalanine to tyrosine content [147] is the only protein for which phenylalanine contribution to the emission is reported [146,148]. This assignment is made on the basis of phosphorescence fine structure and lifetimes as well as the value of the rms zero-field parameter, D', determined from the $\Delta m_s = 2$ EPR signal found in collagen [148]. It appears to us, however, that incorrect zero-field parameters were quoted in the arguments given in Ref. [148] (compare with Ref. [70]) and we suggest that the collagen EPR signal probably is due to a small amount of a tryptophan-containing protein contaminating the sample. Furthermore, the wavelengths and the fine structure of the longer component of the decay of the collagen phosphorescence are also more nearly characteristic of tryptophan than of phenylalanine. Finally, although it is difficult to quantify this point, it seems that the exciting source used [148] would result in almost exclusive excitation of tyrosine (and any tryptophan present) so it is difficult to see how 10% of the phosphorescence intensity which contributes to the long-lived decay can be attributed to phenylalanine.

The phosphorescence spectra from Class A proteins are frequently red shifted from that of free tyrosine and have significantly shorter lifetimes and reduced quantum yields. Although the incorporation of tyrosine into a polypeptide chain may partially account for these observations, the full explanation appears to lie frequently in the proximity of the emitting residues to disulfide or sulfhydryl groups [114]. Effects associated with neighboring disulfide linkages appear to be quite common in Class A proteins (e.g., ribonuclease A) and the phosphorescence spectra in these cases as shown in Fig. 12 closely resemble that of oxytocin [111] a small cyclic polypeptide whose only aromatic residue is a tyrosine which is adjacent to a disulfide bridge.

The triplet lifetimes at 77 °K of proteins showing an oxytocin-type phosphorescence are in the range of 1.4—1.7 sec [28b,149], similar to the value 1.7 sec found from oxytocin [111], whereas the lifetime of the tyrosine zwitterion itself is $\sim$2.9 sec [28b]. The phosphorescence lifetime of the dipeptide dimer cystinyl-bis-tyrosine also is 1.7 sec [28b] in contrast with the 2.6 sec lifetime found for the reduced dipeptide cysteinyl-tyrosine [149]. Thus, the shortened lifetime is due to the neighboring disulfide linkages and accordingly when the disulfide bridges are reduced the observed protein or peptide lifetimes are found to increase [28b]. The ribonuclease A phosphorescence is reported to decay as two first order processes at 77 °K with decay lifetimes of 0.5 sec and 1.6 sec [30]. The reduced tyrosine phosphorescence lifetime may involve electron transfer to disulfide from the triplet [115].

The UV destruction of cystine bonds in enzymes, and loss of enzymatic activity has been associated recently with the presence of tyrosine. Ultraviolet (254 nm) destruction of the enzymatic activity of duck lysozyme II correlates well with the destruction of cystine, whereas little tryptophan destruction is observed. In hen lysozyme, which contains two tyrosines fewer than the duck enzyme, comparatively less destruction of cystine, and more destruction of tryptophan is associated with the UV loss of enzymatic activity [150].

Other vicinal groups which may be involved in quenching of the tyrosine emission are amide and peptide links [41], α-amino groups [40],

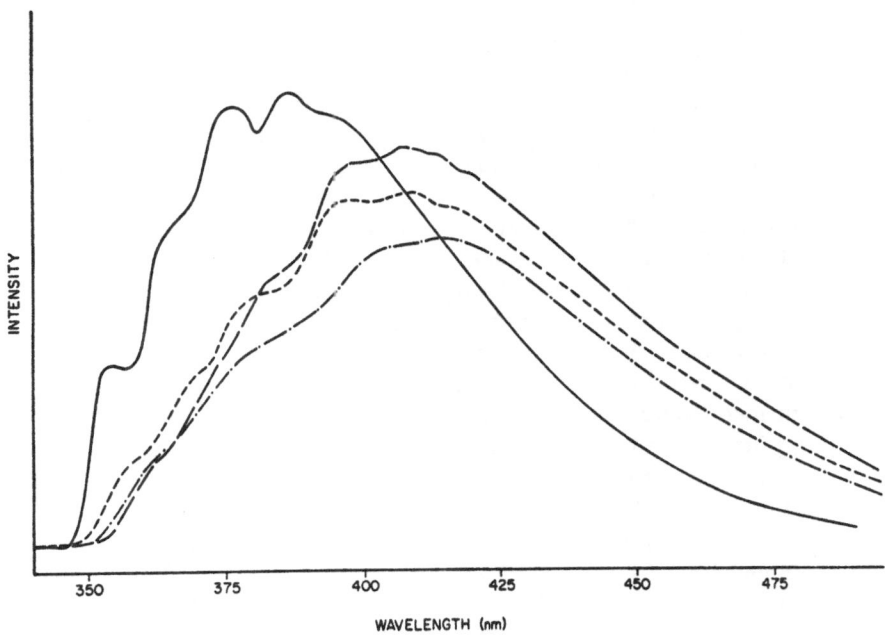

Fig. 12. Phosphorescence of ribonuclease A (a Class A protein). Solvent is ethylene glycol-water, pH 7; temperature is 77 °K with excitation at 278 nm. (————) N-acetyl-L-tyrosine amide; (-------) oxytocin; (———) ribonuclease A; (-··-··-) L-cystinyl-bis-L-tyrosine. (From Longworth [111]).
[From Longworth: Excited states of proteins and nucleic acids, pp. 319—484. Fig. 37, p. 387 (1971).]

lysine and ammonium [112,151], carboxylate groups including glutamate [112,152] and aspartate as well as the corresponding carboxylamides and other carbonyl derivatives [153,154].

The low quantum yields of Class A protein phosphorescence are partially restored to those of tyrosine by denaturing the protein. Reduction of the disulfide linkages in addition to denaturation usually results in luminescence yields, lifetimes and Stokes' shifts nearly identical to those of tyrosine.

The features of the tyrosine contribution to the fluorescence of Class B proteins are no different from those found in Class A proteins [155]. Again, though, there are minor variations in the wavelengths of the phosphorescence maxima and variations in the yields and lifetimes indicative of neighboring disulfide bridges and other quenching groups [111,156]. The only additional factor which may be present in Class B proteins is the possibility of tyr→trp transfer. A large number of reports examining this problem have been referenced in the texts of Konev [8] and Longworth [28b]. Certainly on the basis of observations of the aromatic amino acids free in solution and incorporated into polypeptides, tyr→trp transfer might be expected to be quite efficient in some proteins. The quantum yields of tyrosine fluorescence and phosphorescence of Class B proteins, however, although much lower than for free tyrosine, are generally similar to those found in Class A proteins. Thus tyr→trp transfer apparently is not as effective in the quenching of the tyrosine emission of many proteins as is the quenching by other vicinal groups. Indeed the phosphorescence excitation difference spectra technique only yielded evidence of tyr→trp transfer in two of seven enzymes with high tyr/trp content [157]. Several proteins for which tyr→trp transfer has been quantitatively estimated by monitoring changes in tryptophan quantum yields with varying exciting wavelength are: ribonuclease T_1 where 30—50% of the energy absorbed by tyrosine is transferred to tryptophan [156,158], pepsin —25% [159], papain —56% [160] B.subtilis-α-amylase-15% [157], and alcohol dehydrogenase —32% [157].

At alkaline pH the fluorescence of Class B proteins is found to be that of tyrosinate [161–163]. A report of Vladimirov and Zimina [155] that the fluorescence of serum and egg albumins at pH 13 is entirely due to tryptophan is probably in error; the observed luminescence is most likely that of tyrosinate. Tryptophan and tyrosinate fluorescence spectra are quite similar and lifetime measurements are sometimes necessary for definite identification. The phosphorescence of Class B proteins at alkaline pH generally has a considerable tryptophan component along with a dominant tyrosinate contribution [26,164]. Thus, trp→tyr$^{(-)}$ transfer appears to be very efficient at the singlet level and enhanced intersystem crossing to the tryptophan triplet at high pH [26] also contributes to the tryptophan fluorescence quenching and to the production of tryptophan phosphorescence of Class B proteins at high pH. It is possible that tyr$^{(-)}$→trp triplet transfer also occurs to an extent in some proteins.

Tabulations of the phosphorescence properties of a large number of Class B proteins are available in the literature [8,28b]. Peak wavelengths of the 0—0 bands of tryptophan vary over a range of ~10 nm, the extreme cases being ribonuclease T_1 (406 nm) which coincides with that found for free tryptophan, and lysozyme, which has a 0—0 band at 416 nm. The other peaks of protein phosphorescence spectra are similarly red shifted from tryptophan in solution. Bandwidths of the protein phosphorescence peaks due to tryptophan generally are on the order of ~200 cm^{-1}, again similar to those of the monomer. Several exceptions which show somewhat narrower ~60 cm^{-1} bandwidths and more clearly resolved fine structure are ribonuclease T_1, elastase, avidin and subtilisin carlsberg.

Denaturation of the protein results in blue shifts of the phosphorescence to match the emission of tryptophan in a similar medium. In the instances of narrow bandwiths mentioned above, denaturation broadens the phosphorescence peaks to the more commonly found 200 cm^{-1}. Thus, the variations noted in the phosphorescence spectra of Class B proteins appear to be due to differing amounts of shielding of the emitting chromophores from the solvent medium. The variations are for the most part removed with denaturation when all tryptophan residues are given roughly equal exposures to the solvent.

The phosphorescence quantum yields of several Class B proteins approach that of tryptophan in frozen glasses, but more commonly they are considerably less than this value. As with tyrosine, there are a number of innate groups in the protein which may contribute to tryptophan quenching. The more common of these are: carboxylate groups — glutamate [113]), aspartate [44]), and the α-carboxyl of the C-terminus of the peptide chain [165–167]); disulfide linkages [114]), sulfhydryl groups [168–169]), the α-amino of the N-terminus of the peptide chain [166,167,170]) and the imidazolium cation of histidine [108,171]). Again, as with tyrosine the quenching may be partially lifted by denaturation and reduction of disulfide linkages.

Protein phosphorescence lifetimes are generally within the range 5.5—6.5 sec, the decay being a single exponential at 77 °K except when a faster-decaying tyrosine component also is observable. A notable exception is lysozyme where a relatively short non-exponential decay is found in spite of the absence of any observable tyrosine contribution to the emission [116,172]). Lifetimes of 1.4 and 4.4 sec are reported at 77 °K with 5/6 of the intensity decaying with the shorter lifetime [114,172]). The same authors found also that denaturation and reduction of disulfide links yields an exponential decay with a 5.6 sec lifetime. More recently it has been observed that the lysozyme phosphorescence decay at 4.2 °K reverts to a single exponential with a lifetime of 6.3 sec [173]). Thus, it seems likely that the short lifetime at 77 °K is due to a thermally activated process of some sort which may well be a transfer of an electron from a triplet state to a neighboring disulfide linkage (similar to that found for tyrosine [115])). At 4.2 °K the excited triplet state is trapped at the tryptophan sites possibly because side chain motion with an activation barrier is involved in electron transfer. It should be mentioned that the three-dimensional structure of lysozyme is known from the X-ray determination of Phillips and co-workers [174] and five of the six tryptophan residues have neighboring disulfide links. In any event, the explanation for the reduced triplet lifetime in terms of a simple external heavy atom effect caused by disulfide linkages appears to be ruled out by the normal phosphorescence decay at 4.2 °K. It might be interesting to determine whether the tyrosine phosphorescence lifetime from oxytocin-type Class A proteins is temperature-dependent below 77 °K.

The sensitivity of tryptophan phosphorescence as a probe for distinguishing protein sites is largely lost because the breadth of the emission from any single site generally is sufficient to mask any shifts between the various sites. Nonetheless several cases have been found where the emission from distinct sites is optically resolvable [175–177]. The most striking example is the enzyme, horse liver alcohol dehydrogenase (HLAD) whose phosphorescence spectrum is composed of two apparent tryptophan emissions with origins shifted from each other by ~300 cm^{-1} [176] (see Fig. 13). By differential quenching, sensitization studies, and heavy atom solvent effects, Purkey and Galley [176] convincingly

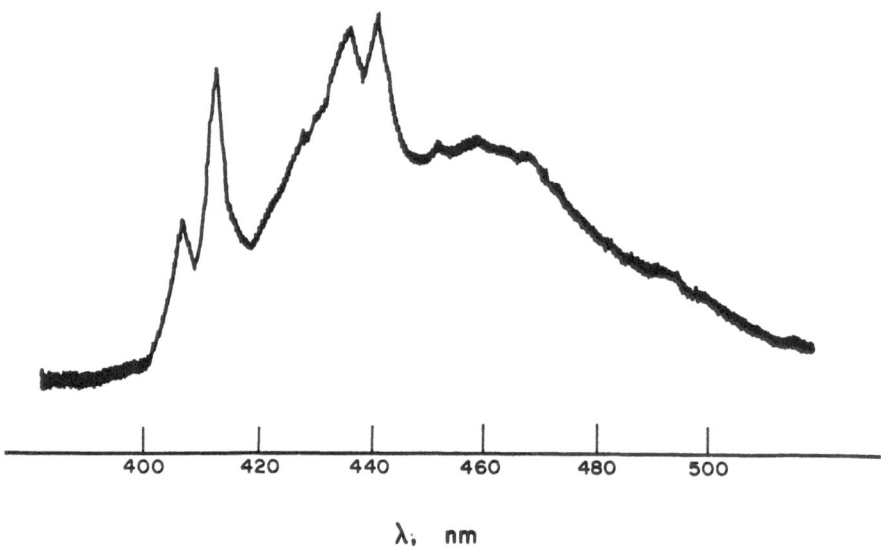

400   420   440   460   480   500

λ, nm

Fig. 13. Phosphorescence emission of the enzyme horse liver alcohol dehydrogenase at 1.3 °K in 50 % ethylene glycol-water showing the optically-resolved origins of the two tryptophan sites of the enzyme. Excitation is at 295 nm. (From Zuclich et al. [101])

showed that the blue-shifted emission was due to a tryptophan site on the surface of the protein, fully exposed to the solvent, while the red-shifted emission was attributed to a tryptophan site buried within the protein in a less polar but more polarizable medium than the aqueous solvent. HLAD is an enzyme of molecular weight 80,000 composed of two identical subunits each containing only two tryptophans [178]. The structure of HLAD has been determined recently at a resolution of 2.9 Å [179]. Tryptophan No. 15 is nearly completely exposed to the solvent whereas trp No. 314 is shielded from the solvent since it lies in the contact area of the monomers in the dimer structure.

In several cases where there is no sign of resolution in the optical emission, differential quenching techniques have been used to indicate heterogeneity. This has been done by monitoring changes in shape and peak wavelengths of emission upon the binding of a substrate or substrate analog to an enzyme [180–184] or upon the binding of dyes [185] or chromophoric inhibitors which quench the protein emission by energy transfer [19, 159,186]. The problems inherent in these techniques are in assessing to what degree the changes in the emission are due to the following:

(1) direct interactions of the bound molecule with the tryptophan residues,

(2) shielding of the emitting residues from the solvent or other interacting vicinal groups in the presence of the bound molecule, and,

(3) minor conformational changes in the protein which may be caused by binding a foreign molecule.

4.2. EPR and ODMR of Proteins

The early reports of EPR measurements on triplet states of the aromatic amino acids included the results on several proteins [66,187]. These studies, which were restricted to $\Delta m_s = 2$ signals, did not note any difference between the protein EPR signals and those of the free aromatic amino acids. Later studies, which included the measurement of $\Delta m_s = 1$ signals, also failed to reveal any significant variations between protein EPR signals and those of free tryptophan and tyrosine [70,73]. Thus, conventional triplet state EPR spectroscopy has failed to yield any new information concerning protein triplet states.

As we have described earlier, the ODMR method can be used to determine the radiative pattern of the triplet sublevels as well as the individual sublevel decay constants, the relative intersystem crossing rates, and spin-lattice relaxation rates. To the extent that these kinetic parameters (as well as the zero-field splittings) are sensitive to the environment of the triplet chromophore, ODMR appears to be considerably

151

more promising as a probe of protein structure than either conventional EPR spectroscopy, or luminescence spectroscopy, since it combines elements of both.

Thus far, ODMR measurements have been reported [99–101,173] only on the proteins bovine serum albumin, HLAD, lysozyme, and lysozyme complexed with the inhibitor tri-N-acetylglucosamine (tri-NAG), and in these proteins only the tryptophan ODMR has been studied in detail.

The tryptophan phosphorescence spectrum of these proteins is inhomogeneously broadened. The observed spectrum is a superposition of emissions originating from a distribution of microenvironments in the sample. It is thus possible to optically select a portion of the distribution by observing the phosphorescence through narrow monochromator slits. ODMR measurements then can be made on the wavelength-selected portion of the emitting population. A correlation between the triplet energy (determined by the selected wavelength) and the zero-field splitting parameters $|D|$ and $|E|$ thus could be revealed by ODMR measurements. Such a correlation is found for the tryptophan triplet in ethylene glycol–H_2O and is shown in Fig. 14. $|D|$ and $|E|$ vary in what appears to be a linear manner with the wavelength of observation throughout the 0–0 band. This experiment reveals the inhomogeneous nature of the tryptophan phosphorescence emission. The ODMR linewidth also increases when the measurements are made on populations which emit away from the center of the 0–0 band (Fig. 14).

The contribution of more than one type of tryptophan site to the phosphorescence is revealed by a discontinuity in the plot of $|D|$ and $|E|$ vs. the detected wavelength in certain proteins. Fig. 15 shows a plot of $|D|$ and $|E|$ vs. λ for HLAD, lysozyme, and the lysozyme-tri-NAG complex. The discontinuity for HLAD occurs at the wavelength at which the 0–0 bands of the optically resolved tryptophans overlap (see Fig. 13). Apparently the $|D|$ and $|E|$ values of tryptophans in distinct protein sites correlate differently with λ resulting in a discontinuity in the region where the emissions from the distinct sites overlap. It can be seen that the data for the blue-shifted tryptophan site in HLAD (HLAD-1) lie very close to those of tryptophan in ethylene glycol–H_2O confirming the assignment of this emission to a solvent-exposed tryptophan [176]. Although separate 0–0 bands from distinct sites are not resolved in the phosphorescence of lysozyme, it is apparent from the large discontinuities in Fig. 15 that more than one tryptophan site contributes to the emission. The behavior of the lysozyme-tri-NAG complex is very similar to that of lysozyme; only small differences in the $|D|$ and $|E|$ values are observed. The behavior of crystalline samples of lysozyme, and the lysozyme tri-NAG complex are the same as those of the ethylene glycol–H_2O

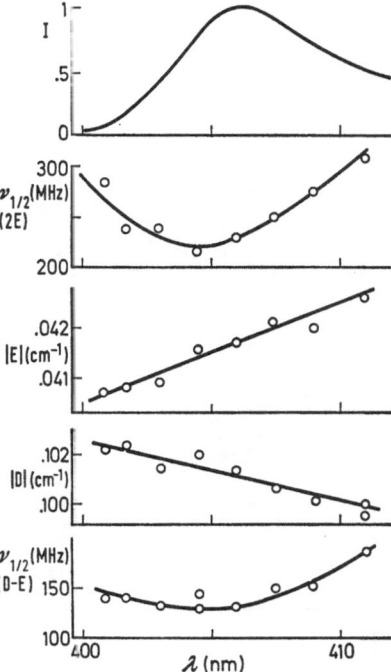

Fig. 14. Zero-zero phosphorescence emission band of tryptophan in ethylene glycol-water at 1.25 °K (top). Below this, in order, are shown the half-width of the zero-field $|2E|$ magnetic resonance transition, the zero-field parameter $|E|$, the zero-field parameter $|D|$, and the half-width of the zero-field $|D|-|E|$ transition, each as a function of the monitored optical wavelength with narrow monochromator slits. (From von Schütz et al. [173])

samples, except that the ODMR linewidths are far greater in the crystalline samples. Recent measurements [182,188] suggest that the room temperature fluorescence of lysozyme originates predominantly from tryptophan 108 near the catalytic site with a smaller contribution from tryptophans 62 and 63 also at the active site. Since tryptophans 62 and 63 are hydrogen bonded to the tri-NAG inhibitor, the small effect of inhibitor binding observed in the ODMR spectra suggests that these tryptophans do not contribute to the low temperature phosphorescence of lysozyme. It is possible, of course, that at low temperature the luminescence might originate mainly from the three remaining tryptophan residues which are not in contact with the catalytic site.

One other peculiarity of the ODMR properties of lysozyme should be mentioned. Whereas the ODMR lines of tryptophan and HLAD are

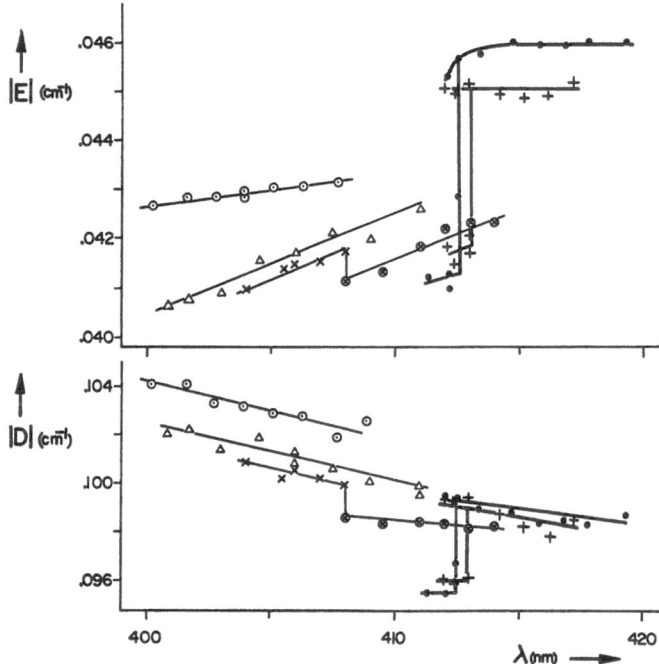

Fig. 15. Zero-field splitting parameters $|D|$ and $|E|$ obtained by ODMR as a function of phosphorescence wavelength monitored with narrow slits in the 0—0 band region of indole, and tryptophan-containing samples plotted against monitored wavelength. The solvent is 50% ethylene glycol—water and $T = 1.25\,^\circ$K. Indole, $\odot$; tryptophan$\triangle$; HLAD-1, x; HLAD-2, $\otimes$; lysozyme $\cdot$; and lysozyme-tri-NAG complex, $+$. (From von Schütz et al.[173])

found to be inhomogeneously broadened and it is possible to burn 10 MHz wide holes in them by application of monochromatic microwave power, the ODMR lines of lysozyme (although $\sim$100 MHz wide) behave as homogeneously broadened lines [102,173]. Attempts to burn holes in lysozyme ODMR lines result in the saturation of the entire line at sufficiently high microwave power. Such behavior is consistent with the presence (at 1.2 $^\circ$K) of an averaging process operative over the entire ODMR linewidth. An averaging process also is consistent with the apparent *absence* of wavelength dependence of $|D|$ and $|E|$ in lysozyme (Fig. 15). Such a wavelength-dependence is a characteristic of each of the other samples investigated to date. The averaging process does not occur, however, between the distinct emitting sites of lysozyme which are resolved in Fig. 15. We do not understand the nature of this averaging

154

process, but it might involve interactions between proximate tryptophan chromophores.

It was mentioned earlier that the linewidth of the ODMR signal increases as the phosphorescence is monitored towards the edges of the 0—0 emission band. If the emission is monitored at the center of the 0—0 band with narrow monochromator slits, a trend in the linewidth of the ODMR signals is also observed between samples. It appears that an increase in the ODMR linewidth is correlated with a blue shift in the phosphorescence origin as shown in Fig. 16 [173]. This increase in line-width may be connected with increasing exposure of the tryptophan site to the solvent medium. The $|2E|$ ODMR transition is always observed

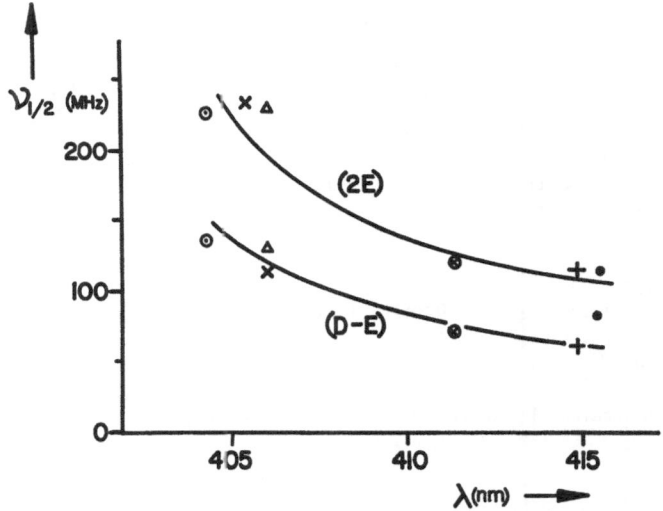

Fig. 16' Linewidths of the ODMR $|2E|$ and $|D|—|E|$ transitions monitored at the maximum of the 0—0 emission band plotted *vs.* peak wavelength. $T = 1.25\,°K$, and the solvent is 50% ethylene glycol-H_2O. Indole, ⊙; HLAD-1, x; tryptophan, △; HLAD-2, ⊗; lysozyme, · ; and lysozyme-tri-NAG complex, +. (From von Schütz *et al.* [173])

to be wider than the transition which occurs at $|D|—|E|$, perhaps indi-cating that E is more sensitive to environmental perturbations than D.

The only protein for which measurements of the individual triplet sublevel decay constants and spin-lattice relaxation rate constants have been measured is lysozyme, and these only for the major emitting tryptophan sites (narrow-band optical detection at 415 nm). Measure-

ments were made by the method described in Section 2.3.2. in which
the phosphorescence decay is observed during microwave saturation of
pairs of sublevels. The results along with the spin-lattice relaxation
rate constants are compared with those obtained from similar measurements on tryptophan in ethylene glycol-H_2O glass in Table 6. The decay
constants are similar, and in the order $k_x > k_y > k_z$. (See Fig. 8). Only the
T_x level is radiative in each case. Spin-lattice relaxation in lysozyme
occurs predominantly between the T_y and T_z sublevels only, whereas

Table 6. Sublevel decay constants and spin-lattice relaxation rates of the tryptophan
triplet state of hen lysozyme compared with tryptophan Zwitterion [99]

Sample	Total sublevel decay rates			Spin-lattice relaxation rates[1]		
	k_x (sec^{-1})	k_y (sec^{-1})	k_z (sec^{-1})	W_z (sec^{-1})	W_y (sec^{-1})	W_x (sec^{-1})
Tryptophan (ethylene glycol—H_2O) excitation, 295 nm detection, 406.0 nm	0.240	0.119	0.038	−0.002	0.036	0.036
Lysozyme (ethylene glycol—H_2O) excitation, 298 nm detection, 415.0 nm	0.293	0.113	0.054	−0.003	0.006	0.084

[1] Small negative W's are zero within the estimated experimental accuracy of $\pm$.005

comparable spin-lattice relaxation occurs between T_z and both T_y and
T_x in tryptophan. Relaxation between T_x and T_y is slow in each case.
It is not presently possible to rationalize this difference in spin-lattice
pathways in terms of any model of the interaction of the chromophore
with its environment. It should be pointed out though that spin-lattice
relaxation rates between the triplet sublevels of tryptophan appear
to be one of the more environment-sensitive parameters which have
been obtained from ODMR methods. Clearly, more proteins and polypeptides should be measured in order to determine whether differences
in spin-lattice relaxation patterns are, in fact, general and can be correlated with specific types of interactions.

The individual sublevel decay constants may be used according to
Eq. (1) to predict the average triplet state decay lifetime which would
occur for rapid spin-lattice relaxation. The average decay lifetimes are

predicted to be 7.6 sec and 6.5 sec for tryptophan and lysozyme, respectively. The most reliable decay lifetime for the tryptophan Zwitterion at 77 °K in ethylene glycol–H_2O glass appears to be [28b] 6.65 sec (Table 1). The reduction of the 77 °K lifetime below that predicted by the ODMR measurements at 1.25 °K shows that intersystem crossing from the triplet is partially a thermally activated process at 77 °K. The predicted lifetime of the lysozyme triplet is in agreement with the measured [173] phosphorescence decay lifetime of 6.3 sec at 4.2 °K (spin-lattice relaxation in lysozyme at liquid helium temperature is sufficiently rapid to yield a single average exponential decay). The shorter lifetimes at 77 °K (1.4 sec and 4.4 sec) thus are clearly indicative of thermally-activated processes although probably of a more complicated nature than those occurring in tryptophan.

The relative populating rates of the triplet sublevels during optical pumping at 1.2 °K have been measured for tryptophan [99] and for lysozyme [99] (major luminescent component of Fig. 15), using flash excitation followed rapidly (~150 msec delay) by fast-passage microwave saturation of pairs of sublevels [89-92]. Since the sequence takes place in a short time interval compared with sublevel decay and spin-lattice relaxation, the sublevel populations sampled by the microwave saturation give the relative populating rates directly. The results are presented in Table 7. The populating mechanism for the tryptophan

Table 7. Relative populating rates of the triplet sublevels of tryptophan and the tryptophan moieties of hen lysozyme emitting at 415 nm [99]

Sample	Relative populating rates (%)[1]		
	P_x	P_y	P_z
Tryptophan (ethylene glycol–H_2O) excitation, 295 nm detection, 406 nm	39	38	23
Hen lysozyme (ethylene glycol–H_2O) excitation, 298 nm detection, 415 nm	39	36	25

[1] Estimated accuracy, $\pm 1\%$, axes shown in Fig. 8.

phosphorescent state is purely intersystem crossing from S_1, since triplet-triplet energy transfer is not probable at the high dilution (10^{-3} M) of tryptophan. The results reveal no selectivity of intersystem crossing between the two in-plane sublevels T_x and T_y (see Fig. 8), whereas the out-of-plane sublevel T_z has a significantly smaller intersystem crossing rate than the in-plane sublevels. Such an intersystem crossing pattern is expected for $^3\pi,\pi^*$ states [95]. In lysozyme, there is a similar pattern in the triplet populating mechanism. Triplet-triplet energy transfer between tryptophan sites in lysozyme could occur because of the relatively high density of tryptophans in this enzyme. If the tryptophans are translationally inequivalent, a loss of spin alignment is a consequence [93,189]. The agreement between the relative sublevel populating rates in tryptophan and lysozyme makes it unlikely that triplet-triplet transfer is an important populating mechanism for the lysozyme triplets emitting at 415 nm (the wavelength at which the phosphorescence was monitored).

The ODMR measurements of aromatic amino acid and protein triplet states are very recent and in this review we have tended to emphasize this technique because it is in its infancy and relatively unfamiliar. Also, we think that it shows promise as a powerful general method for studying the detailed nature of the excited states and of energy transfer processes in proteins and other biopolymers. Interestingly enough, the necessity of making ODMR measurements at temperatures in the liquid helium range lead to the discovery that thermally-activated processes still affect the triplet states of lysozyme at 77 °K. Further discoveries of this nature may be forthcoming as other familiar biopolymers are studied at very low temperatures. Proteins may be quite dynamic molecules at 77 °K, at least as far as their excited triplet states are concerned.

Acknowledgments: This work was supported in part by the National Institutes of Health, U.S. Public Health Service, through postdoctoral a fellowship (JZ) and a research grant (AHM). We also received research grant support from the National Science Foundation. We wish to thank the copyright holders for permission to use certain of the figures.

5. References

[1] Becconi, J. B.: Phil. Trans. *44*, 81 (1746) (English Transl.).

[2] Hoshijima, S.: Sci. Pap. Inst. Phys. Chem. Res., Tokyo *20*, 109 (1933).

[3] Leighton, W. G., Leighton, P. A.: Can. J. Chem. *12*, 139 (1935).

[4] Giese, A. C., Leighton, P. A.: Science *85*, 428 (1937).

[5] Vladimirov, Yu. A., Litvin, F. F.: Biofizika *4*, 601 (1959).

[6] Konev, S. V., Katibnikov, M. A.: Biofizika *6*, 638 (1961).

[7] Katibnikov, M. A., Konev, S. F.: Biofizika *7*, 150 (1962).

8) Konev, S. V.: Fluorescence and phosphorescence of proteins and nucleic acids. New York: Plenum Press 1967.
9) Sapezhinskii, I. I.: Biofizika 10, 429 (1965).
10) Sapezninskii, I. I., Silaev, Yu. V.: Dokl. Akad. Nauk SSSR 162, 691 (1965).
11) Debye, P., Edwards, J. O.: Science 116, 143 (1952).
12) Steele, R. H., Szent-Györgyi, A.: Proc. Natl. Acad. Sci. U.S. 44, 540 (1958).
13) Vladimirov, Yu. A., Litvin, F. F.: Biofizika 5, 127 (1960).
14) Jordan, P.: Naturwissenschaften 26, 693 (1938).
15) Szent-Györgyi, A.: Science 93, 609 (1941).
16) Szent-Györgyi, A.: Biochim. Biophys. Acta 16, 167 (1955).
17) Debye, P., Edwards, J. O.: J. Chem. Phys. 20, 236 (1952).
18) Piette, L. H., Rabold, G. P.: In: Magnetic resonance in biological systems (eds. A. Ehrenberg, B. G. Malmstrom and T. Vanngard). Oxford Pergamon Press 1967.
19) Galley, W. C., Stryer, L.: Proc. Natl. Acad. Sci. U.S. 60, 108 (1968).
20) McCarville, M., Hauxwell, R.: Biochim. Biophys. Acta 251, 285 (1971).
21) Kuntz, I. D.: J. Am. Chem. Soc. 93, 514, 516 (1971).
22) Strickland, E. H., Horwitz, J., Billups, C.: Biochemistry 8, 3205 (1969).
23) Yu, N. T., Jo, B. H.: Arch. Biochem. Biophys. 156, 469 (1973).
24) Douzou, P., Sireix, R., Travers, F.: Proc. Natl. Acad. Sci. U.S. 66, 787 (1970).
25) Wetlaufer, D. B.: Advan. Protein Chem. 17, 303 (1962).
26) Truong, T., Bersohn, R., Brumer, P., Luk, C. K., Tao, T.: J. Biol. Chem. 242, 2979 (1967).
27) Teale, F. W. J., Weber, G.: Biochem. J. 65, 476 (1957).
28a) Weinryb, I., Steiner, R. F.: Excited states of proteins and nucleic acids (eds. R. F. Steiner and I. Weinryb), p. 277. New York: Plenum Press 1971.
28b) Longworth, J. W.: Excited states of proteins and nucleic acids (eds. R. F. Steiner and I. Weinryb), p. 319. New York: Plenum Press 1971.
29) Beaven, G. H.: In: Advances in spectroscopy (ed. H. W. Thompson), Vol. 2, p. 331. New York: Interscience, Inc. 1961.
30) Nag-Chaudhuri, J., Augenstein, L.: Biopolymers, Symp. 1, 441 (1964).
31) Bishai, F., Kuntz, E., Augenstein, L.: Biochim. Biophys. Acta 140, 381 (1967).
32) Volotovskii, I. D., Konev, S. V., Chernitskii, Ye. A.: Biofizika 12, 421 (1967).
33) Chernitskii, Ye. A., Konev, S. V., Bobrovich, V. P.: Dokl. Akad. Nauk, BSSR 7, 628 (1963).
34) Steen, H. B.: Photochem. Photobiol. 6, 805 (1967).
35) Van Duuren, B. L.: J. Org. Chem. 26, 2954 (1961).
36) Doub, L., Vendenbelt, J. M.: J. Am. Chem. Soc. 69, 2714 (1947).
37) Campbell, T. W., Linden, S., Godshalk, S., Young, W. G.: J. Am. Chem. Soc. 69, 880 (1947).
38) Barenboim, G. M., Domanskii, A. N., Turoverov, K. K.: Luminescence of biopolymers and cells. New York: Plenum Press 1969.
39) Weinryb, I., Steiner, R. F.: Biochemistry 7, 2488 (1968).
40) White, A.: Biochem. J. 71, 217 (1959).
41) Cowgill, R. W.: Arch. Biochem. Biophys. 100, 36 (1963).
42) Feitelson, J.: J. Phys. Chem. 68, 391 (1964).
43) Rau, H., Augenstein, L.: J. Chem. Phys. 46, 1773 (1967).
44) Cowgill, R. W.: Biochem. Biophys. Acta 200, 18 (1970).
45) Straub, F. B.: Byokemia, Budapest 1961, quoted in Ref. 38), p. 19.
46) McGlynn, S. P., Azumi, T., Kinoshita, M.: Molecular spectroscopy of the triplet state. Englewood Cliffs, N. J.: Prentice Hall, Inc. 1969.
47) Chernitskii, Ye. A., Konev, S. V.: Dokl. Akad. Nauk BSSR 8, 258 (1964).

48) Chignell, D. A., Gratzer, W. B.: J. Phys. Chem. 72, 2934 (1968).
49) Carter, J. G., Nelson, D. R., Augenstein, L. G.: Arch. Biochem. Biophys. 111, 270 (1965).
50) Bogach, P. G., Zima, V. L., Filenko, A. M.: Biofizika 16, 340 (1970).
51) Grossweiner, L. I., Swenson, G. W., Zwicker, E. F.: Science 141, 805 (1963).
52) Vladimirov, Yu. A.: Photochemistry and luminescence of proteins, Nauka, Moskow 1965; Israel Program for Scientific Translations, Jerusalem 1969.
53) Koudelka, J., Augenstein, L.: Photochem. Photobiol. 7, 613 (1968).
54) Kuntz, E.: Nature 217, 845 (1968).
55) van der Waals, J. H., de Groot, M. S.: Mol. Phys. 2, 333 (1959).
56) de Groot, M. S., van der Waals, J. H.: Mol. Phys. 3, 190 (1960).
57) Hutchison, C. A., Mangum, B. W.: J. Chem. Phys. 29, 952 (1958).
58) Hutchison, C. A., Mangum, B. W.: J. Chem. Phys. 34, 908 (1961).
59) Yager, W. A., Wasserman, E., Cramer, R. M. R.: J. Chem. Phys. 37, 1148 (1962).
60) Smaller, B., Remko, J.: Organic Crystal Symposium. National Research Council, Ottawa, Canada 1962.
61) Wasserman, E., Snyder, L., Yager, W. A.: J. Chem. Phys. 41, 1763 (1964).
62) Kottis, Ph., Lefebvre, R.: J. Chem. Phys. 41, 379 (1964).
63) Smaller, B.: Nature 195, 593 (1962).
64) Ptak, M., Douzou, P.: Nature 199, 1092 (1963).
65) Smaller, B.: Advan. Biol. Med. Phys. 9, 225 (1963).
66) Shiga, T., Piette, L. H.: Photochem. Photobiol. 3, 223 (1964).
67) Maling, J. E., Rosenheck, K., Weissbluth, M.: Photochem. Photobiol. 4, 241 (1965).
68) Cailly, C., Boukhors, A.: Compt. Rend. Acad. Sci. Paris 264C, 480 (1967).
69) ten Bosch, J. J., Rahn, R. O., Longworth, J. W., Shulman, R. G.: Proc. Natl. Acad. Sci. U.S. 59, 1003 (1968).
70) Zuclich, J.: J. Chem. Phys. 52, 3586 (1970).
71) de Groot, M. S., Hesselman, I. A. M., van der Waals, J. H.: Mol. Phys. 16, 45 (1969).
72) Smaller, B., Avery, E. C., Remko, J. R.: J. Chem. Phys. 46, 3976 (1967).
73) Zuclich, J.: Ph. D. Thesis, Columbia University (1969).
74) Gribova, Z. P.: In: Elektrochemische Methoden und Prinzipien in der Molekular-Biologie. III. Jenaer Symposium (ed. H. Berg). Berlin: Akademie-Verlag 1966.
75) Azizova, O. A.: In: Elektrochemische Methoden und Prinzipien in der Molekular-Biologie. III. Jenaer Symposium (ed. H. Berg). Berlin: Akademie-Verlag 1966.
76) Azizova, O. A., Gribova, Z. P., Kayushin, L. P., Pulatova, M. K.: Photochem. Photobiol. 5, 763 (1966).
77) Siegel, S., Eisenthal, K.: J. Chem. Phys. 42, 2494 (1965).
78) Steen, H. B., Photochem. Photobiol. 9, 479 (1969).
79) Santus, R., Hélène, C., Ptak, M.: Photochem. Photobiol. 7, 341 (1968).
80) Azizova, O. A.: Biofizika 9, 745 (1964).
81) Hélène, C., Santus, R., Douzou, P.: Photochem. Photobiol. 5, 127 (1966).
82) Lhoste J. M., Hélène, C., Ptak, M., In: The triplet state (ed. A. B. Zahlan), p. 479. Cambridge: Cambridge Press 1967.
83) El-Sayed, M. A., Siegel, S.: J. Chem. Phys. 44, 1416 (1966).
84) Lhoste, J. M., Haug, A., Ptak, M.: J. Chem. Phys. 44, 648, 654 (1966).
85) Zuclich, J.: J. Chem. Phys. 52, 3592 (1970).
86) Sharnoff, M.: J. Chem. Phys. 46, 3263 (1967).

160

87) Kwiram, A.: Chem. Phys. Letters *1*, 272 (1967).
88) Schmidt, J., Hesselmann, I. A. M., de Groot, M. S., van der Waals, J. H.: Chem. Phys. Letters *1*, 434 (1967).
89) Schmidt, J., Antheunis, D. A., van der Waals, J. H.: Mol. Phys. *22*, 1 (1971).
90) Burland, D. M., Schmidt, J.: Mol. Phys. *22*, 19 (1971).
91) Harris, C. B., Hoover, R. J.: Chem. Phys. Letters *12*, 75 (1971).
92) Harris, C. B.: J. Chem. Phys. *54*, 972 (1971).
93) Winscom, C. J., Maki, A. H.: Chem. Phys. Letters *12*, 264 (1971).
94) Tinti, D. S., El-Sayed, M. A., Maki, A. H., Harris, C. B.: Chem. Phys. Letters *3*, 343 (1969).
95) van der Waals, J. H., de Groot, M. S.: In: The triplet state (ed. A. B. Zahlan), p. 101. Cambridge: Cambridge Press 1967.
96) Schmidt, J., Veeman, W. S., van der Waals, J. H.: Chem. Phys. Letters *4*, 34 (1969).
97) Antheunis, D. A., Schmidt, J., van der Waals, J. H.: Chem. Phys. Letters *6*, 255 (1970).
98) Harris, C. B., Hoover, R. J.: J. Chem. Phys. *56*, 2199 (1972).
99) Zuclich, J., von Schütz, J. U., Maki, A. H.: Mol. Phys. *28*, 33 (1974).
100a) Kwiram, A. L.: In: MTP International Rev. of Science, Phys. Chem., Series One (ed. C. A. McDowell), Vol. 4, p. 271. London: Butterworths 1972.
100b) Zuclich, J., Schweitzer, D., Maki, A. H.: Biochem. Biophys. Res. Commun. *46*, 1764 (1972).
101) Zuclich, J., Schweitzer, D., Maki, A. H.: Photochem. Photobiol. *18*, 161 (1973).
102) Zuclich, J., von Schütz, J. U., Maki, A. H.: J. Am. Chem. Soc. *96*, 710 (1974).
103) Shpol'skii, E. V.: Soviet Phys.-Usp. (Engl. Transl.) *6*, 411 (1963).
104) Kuan, T. S., Tinti, D. S., El-Sayed, M. A.: Chem. Phys. Letters *4*, 507 (1970).
105) Leung, M., El-Sayed, M. A.: Chem. Phys. Letters *16*, 454 (1972).
106) Teale, F. W. J.: Biochem. J. *76*, 381 (1960).
107) Eisinger, J.: Biochemistry *8*, 3902 (1969).
108) Shinitzky, M., Goldman, R.: European J. Biochem. *3*, 139 (1967).
109) Edelhoch, H., Lippoldt, R. E.: J. Biol. Chem. *244*, 3876 (1968).
110) Cowgill, R. W.: Arch. Biochem. Biophys. *104*, 84 (1964).
111) Longworth, J. W.: Photochem. Photobiol. *7*, 587 (1968).
112) Fasman, G. D., Norland, K., Pesce, A.: Biopolymers, Symp. *1*, 325 (1964).
113) Fasman, G. D., Bodenheimer, E., Pesce, A.: J. Biol. Chem. *241*, 916 (1966).
114) Cowgill, R. W.: Biochim. Biophys. Acta *140*, 37 (1967).
115) Feitelson, J., Hayon, E.: Photochem. Photobiol. *17*, 265 (1973).
116) Longworth, J. W.: Biopolymers *4*, 1131 (1966).
117) Madison, S. V., Schellman, J.: Biopolymers *9*, 511, 569 (1970).
118) Fasman, G. D.: In: Poly-α-amino acids (ed. G. D. Fasman). New York: M. Dekker, Inc. 1967.
119) Cosani, A., Peggion, E., Verdini, A. S., Terbojevich, M.: Biopolymers *6*, 963 (1968).
120) ten Bosch, J. J., Longworth, J. W., Rahn, R. O.: Biochim. Biophys. Acta *175*, 10 (1969).
121) Auer, H. E., Doty, P.: Biochemistry *5*, 1708 (1966).
122) Lehrer, S. S., Fasman, G. D.: Biochemistry *6*, 757 (1967).
123) Förster, T., Discussions Faraday Soc. *27*, 7 (1959).
124) Weber, G., Teale, F. W. J.: In: The proteins (ed. H. Neurath). New York: Academic Press 1965.

161

125) Weber, G.: In: Fluorescence and phosphorescence analysis (ed. D. M. Hercules). New York: Interscience 1966.
126) Rabinovitch, B.: Arch. Biochem. Biophys. *124*, 258 (1968).
127) Steiner, R. F., Kolinski, R.: Biochemistry 7, 1014 (1968).
128) Hélène, C., Ptak, M., Santus, R.: J. Chim. Phys. *65*, 160 (1968).
129) Kearns, D. R., Case, W. A.: J. Am. Chem. Soc. *88*, 5087 (1966).
130) Marchetti, A. P., Kearns, D. R.: J. Am. Chem. Soc. *89*, 768 (1967).
131) Cassen, T., Kearns, D. R.: Biochem. Biophys. Res. Commun. *31*, 851 (1968).
132) Mantik, D., Maling, J. E., Weissbluth, M.: Biophys. J. *9*, A 160 (1969).
133) Edelhoch, H., Brand, L., Wilchek, M.: Biochemistry *6*, 547 (1967).
134) Knopp, J. A., Longworth, J. W.: Biochim. Biophys. Acta *154*, 436 (1968).
135) ten Bosch, J. J., Longworth, J. W., Rahn, R. O.: Biochim. Biophys. Acta *175*, 10 (1969).
136) Longworth, J. W., Knopp, J. A., ten Bosch, J. J., Rahn, R. O.: In: Molecular Luminescence (ed. E. C. Lim). New York: Benjamin 1969.
137) Longworth, J. W., Rahn, R. O.: Biochim. Biophys. Acta *147*, 526 (1969).
138) ten Bosch, J. J., Knopp, J. A.: Biochim. Biophys. Acta *188*, 173 (1969).
139) Knopp, J. A., ten Bosch, J. J., Longworth, J. W.: Biochim. Biophys. Acta *188*, 185 (1969).
140) Steiner, R. F.: Biochem. Biophys. Res. Commun. *30*, 502 (1968).
141) Longworth, J. W., Battista, M. D. C.: Photochem. Photobiol. *11*, 207 (1970).
142) Weber, G.: Biochem. J. *79*, 29 P (1961)
143) Konev, S. V., Bobrovich, V. P.: Abstr. Proc. Conf. Biol. Action of Ultraviolet Radiation, Vilna 1964.
144) Augenstein, L. G., Nag-Chandhuri, J.: Nature *203*, 1145 (1964).
145) Longworth, J. W.: Biochem. J. *81*, 23 P (1961).
146) Burshtein, E. A.: Dissertation, Moscow (1964).
147) Steiner, R. F.: The chemical foundations of molecular biology. Princeton: D. van Nostrand 1965.
148) Grimes, M. W., Graber, D. R., Haug, A.: Biochem. Biophys. Res. Commun. *37*, 853 (1969).
149) Wampler, J. W.: Ph. D. Thesis, University of Tennessee 1970.
150) Risi, R., Silva, T., Dose, K.: Photochem. Photobiol. *18*, 475 (1973).
151) Weber, G., Rosenheck, K.: Biopolymers, Symp. *1*, 333 (1964).
152) Rosenheck, K., Weber, G.: Biochem. J. *79*, 29 P (1961).
153) Cowgill, R. W.: Biochim. Biophys. Acta *133*, 6 (1967).
154) Cowgill, R. W.: Biochim. Biophys. Acta *168*, 417 (1968).
155) Vladimirov, Yu. A., Zimina, G. M.: Biokhimiya *30*, 1105 (1966).
156) Longworth, J. W.: Photochem. Photobiol. *8*, 589 (1968).
157) Cassen, T., Kearns, D. R.: Biochim. Biophys. Acta *194*, 203 (1969).
158) Eisinger, J., Navon, G.: J. Chem. Phys. *50*, 2069 (1969).
159) Badley, R. A., Teale, F. W. J.: J. Mol. Biol. *44*, 71 (1969).
160) Weinryb, I., Steiner, R. F.: Biochemistry *9*, 135 (1970).
161) Steiner, R. F., Edelhoch, H.: Biochim. Biophys. Acta *66*, 341 (1963).
162) Steiner, R. F., Edelhoch, H.: Nature *192*, 873 (1961).
163) Steiner, R. F., Edelhoch, H.: J. Biol. Chem. *238*, 925 (1963).
164) Herskovitz, T. T., Sorensen, M.: Biochemistry 7, 2523 (1968).
165) Lehrer, S. S., Fasman, G. D.: Biochem. Biophys. Res. Commun. *23*, 133 (1966).
166) Bridges, J. W., Williams, R. T.: Biochem. J. *107*, 225 (1968).
167) Galley, J. A., Edelman, G. M.: Biochim. Biophys. Acta *60*, 499 (1962).
168) Steiner, R. F., Kirby, E. P.: J. Phys. Chem. *73*, 4130 (1969).
169) Steiner, R. F.: Biochemistry *10*, 771 (1971).

170) Sticks, W., Kolthoff, I. M.: Anal. Chem. *25*, 1050 (1953).
171) Shinitzky, M., Fridkin, M.: European J. Biochem. *9*, 176 (1969).
172) Churchich, J. E.: Biochim. Biophys. Acta *92*, 194 (1964).
173) von Schütz, J. U., Zuclich, J., Maki, A. H.: J. Am. Chem. Soc. *96*, 714 (1974).
174) Blake, C. B., Mair, G. A., North, A. C. T., Phillips, C. D., Sharma, V. R.: Proc. Roy. Soc. (London) *B 167*, 365, 378 (1967).
175) Purkey, R. M., Galley, W. C.: Biophys. J. *10*, 240a (1970).
176) Purkey, R. M., Galley, W. C.: Biochemistry *9*, 3569 (1970).
177) Anderson, S. R., Brunori, M., Weber, G.: Biochemistry *9*, 4723 (1970).
178) Cannon, D. J., McKay, R. H.: Biochem. Biophys. Res. Commun. *33*, 942 (1969).
179) Brändén, C.-I., Eklund, H., Nordström, B., Boiwe, T., Söderlund, G., Zeppezauer, E. Ohlsson, I., Åkeson, Å.: Proc. Natl. Acad. Sci. U.S. *70*, 2439 (1973).
180) Lehrer, S. S., Fasman, G. D.: J. Biol. Chem. *242*, 4644 (1967).
181) Bablouzian, B., Grourke, M., Fasman, G. D.: J. Biol. Chem. *245*, 2081 (1970).
182) Teichberg, I., Sharon, N.: F. E. B. S. Letters *7*, 171 (1970).
183) Sharon, N.: Proc. Roy. Soc. (London) *B 167*, 402 (1967).
184) Suelter, C. H.: Biochemistry *6*, 418 (1967).
185) Elkana, Y.: J. Phys. Chem. *72*, 3654 (1968).
186) Galley, W. C., Stryer, L.: Biochemistry *8*, 1831 (1969).
187) Shiga, T., Mason, H. S., Simo, C.: Biochemistry *5*, 1877 (1966).
188) Imoto, T., Forster, L. S., Rupley, J. A., Tanaka, F.: Proc. Natl. Acad. Sci. U.S. *69*, 1151 (1972).
189) Chan, I. Y., Schmidt, J.: Symp. Faraday Soc. *3*, 156 (1969).

Received January 25, 1974.

H. Meier
**Die Photochemie
der organischen Farbstoffe**

168 Abbildungen. XVI, 471 Seiten
1963 (Organische Chemie in
Einzeldarstellungen, Band 7)
DM 98,–; US $ 40.20
ISBN 3-540-03034-4

Inhaltsübersicht: Die Lichtab-
sorption der Farbstoffe.
Lumineszenz der Farbstoffe. Die
photochemischen Umsetzungen
an organischen Farbstoffen. Der
lichtelektrische Effekt der organischen Farbstoffe. – Spezielle
Reaktionen: Die spektrale Sensi-
bilisierung der photographischen
Schicht. Die Farbstoff-Sensibili-
sierung des Photoeffekts anorga-
nischer Halbleiter. Der photody-
namische Effekt. Der Sehvorgang.
Die Photosynthese. – Anhang:
Zum Problem der Energieüber-
tragung.

A. Schönberg
**Preparative
Organic Photochemistry**

In cooperation with G.O. Schenck,
O.A. Neumüller. Second, com-
pletely revised edition of Präpa-
rative Organische Photochemie.
4 figures and 51 tables
XXIV, 608 pages. 1968
Cloth DM 165,–; US $67.70
ISBN 3-540-04325-X

This monograph is an exhaustive
compilation of photochemical
reactions that are of interest to
the organic chemist. Sufficient
experimental details are given to
make this book a highly useful
manual of photochemical labora-
tory techniques for the student
and the specialist.

O. Stasiw
**Elektronen- und Ionenprozesse
in Ionenkristallen
mit Berücksichtigung
photochemischer Prozesse**

107 Abbildungen. VIII, 307 Seiten
1959 (Struktur und Eigenschaften
der Materie in Einzeldarstellungen,
Band 22)
Gebunden DM 80,–; US $32.80
ISBN 3-540-02475-1

Inhaltsübersicht: Statistik von
Störstellen in Ionenkristallen. –
Fehlerordnungsenergie. – Platz-
wechselvorgänge, Diffusion und
Ionenleitung. – Das Absorptions-
spektrum des idealen Ionengitters.
– Das Absorptionsspektrum von
Ionengittern mit stöchiometrischem
Überschuß der Kationen oder
Anionenkomponente. – Absorp-
tionsspektren von Ionengittern mit
Fremdzusätzen. – Elektronische
Störstellentheorie. – Halbleiter-
prozesse. – Lichtelektrische Lei-
tung. – Photochemische Prozesse
in reinen Ionengittern. – Photochemische Prozesse in Ionengittern
mit Zusätzen. – Photochemische
Prozesse in mechanisch verformten
Kristallen. – Störstellen und Kern-
resonanz. – Anwendung der adia-
batischen Näherung auf Kristalle
mit Störstellen.

Preisänderungen vorbehalten
Prices are subject to change
without notice

**Springer-Verlag
Berlin Heidelberg New York**

Photochemie

6 Abbildungen. 197 Seiten (31 Seiten in Englisch). 1967
(Topics in Current Chemistry / Fortschritte der chemischen Forschung,
Band 7, Heft 3). DM 72,—: US $29.60   ISBN 3-540-03796-9

Inhaltsübersicht: R. Steinmetz: Photochemische Carbocyclo-Additions-
reaktionen. D. Elad: Some Aspects of Photoalkylation Reactions.
M. Pape: Die Photooximierung gesättigter Kohlenwasserstoffe.
E. Fischer: Photochromie und reversible Photoisomerisierung.

Photochemistry

11 figures. 224 pages. 1969 (Topics in Current Chemistry / Fortschritte der chemischen Forschung, Band 13, Heft 2)
DM 65,—: US $26.70   ISBN 3-540-04489-2

From the Contents: J.L.R. Williams: Photochemical Reactions of
Polymers. M.B. Rubin: Photochemistry of o-Quinones and -Diketones.
L.B. Jones, V.K. Jones: Photochemical Reactions of Cycloheptartrienes.
C. v. Sonntag: Strahlenchemie von Alkoholen.
E. Koerner von Gustorf, F.-W. Grevels: Photochemistry of Metal
Carbonyls, Metallocenes, and Olefin Complexes.

Photochemistry

50 figures. IV, 236 pages. 1974 (Topics in Current Chemistry /
Fortschritte der chemischen Forschung, Volume 46)
Cloth DM 68,—: US $27.90   ISBN 3-540-06592-X

From the Contents: J. Michl: Physical Basis of Qualitative MO Arguments in Organic Photochemistry. K.-D. Gundermann: Recent
Advances in Research on the Chemiluminescence of Organic Compounds. W.C. Herndon: Substituent Effects in Photochemical Cycloaddition Reactions. W.-D. Stohrer, P. Jacobs, K.H. Kaiser, G. Wiech,
G. Quinkert: Das sonderbare Verhalten elektronen-angeregter 4-Ring-
Ketone.

Preisänderungen vorbehalten
Prices are subject to change without notice

Springer-Verlag
Berlin Heidelberg New York